THE TECH SET

Building Mobile Library Applications

JASON A. CLARK

facet publishing

Published by Facet Publishing,
7 Ridgmount Street, London WC1E 7AE
www.facetpublishing.co.uk

Facet Publishing is wholly owned by CILIP: the Chartered Institute of Library and Information Professionals.

First published in the USA by ALA TechSource, an imprint of the American Library Association, 2012.
This simultaneous UK edition 2012.

British Library Cataloguing in Publication Data
A catalogue record for this book is available from the British Library.

ISBN 978-1-85604-845-3

Printed and bound in the United Kingdom.

This book is dedicated to Jennifer and Piper, who have endured and supported me as I slogged through the creative process known as writing. To Jennifer, I thank you for your incredible patience, support, and guidance as I found my way. I could not have done it without you. To Piper, I apologize for the lack of car chases, cliffhangers, and mysteries contained herein. It's not that kind of book, sweetie.

CONTENTS

Don't miss this book's companion website!

Turn the page for details.

THE TECH SET® Volumes 11–20 is more than just the book you're holding!

These 10 titles, along with the 10 titles that preceded them, in THE TECH SET® series feature three components:

1. This book
2. Companion web content that provides more details on the topic and keeps you current
3. Author podcasts that will extend your knowledge and give you insight into the author's experience

The companion webpages and podcasts can be found at:

www.alatechsource.org/techset/

On the website, you'll go far beyond the printed pages you're holding and:

- Access author updates that are packed with new advice and recommended resources
- Use the website comments section to interact, ask questions, and share advice with the authors and your LIS peers
- Hear these pros in screencasts, podcasts, and other videos providing great instruction on getting the most out of the latest library technologies

For more information on THE TECH SET® series and the individual titles, visit **www.neal-schuman.com/techset-11-to-20**.

FOREWORD

Mobile devices are now an essential part of people's everyday lives. These new devices such as smartphones, iPhones, iPads, and e-book readers are how library patrons are accessing their information today. *Building Mobile Library Applications* is an essential resource that outlines what it takes to plan, develop, and launch a mobile library application. Readers will learn how to develop iPhone and Android applications for their library, how to mobilize their library catalog, and how to create mobile style sheets. Veteran mobile developer Jason Clark outlines the nuts and bolts of creating both simple and complex mobile websites and designing engaging library apps in this stellar handbook.

The ten new TECH SET volumes are designed to be even more cutting-edge than the original ten. After the first ten were published and we received such positive feedback from librarians who were using the books to implement technology in their libraries as well as train their staff, it seemed that there would be a need for another TECH SET. And I wanted this next set of books to be even more forward-looking and tackle today's hottest technologies, trends, and practices to help libraries stay on the forefront of technology innovation. Librarians have ceased sitting on the sidelines and have become technology leaders in their own right. This series was created to offer guidance and inspiration to all those aspiring to be library technology leaders themselves.

I originally envisioned a series of books that would offer accessible, practical information that would teach librarians not only how to use new technologies as individuals but also how to plan and implement particular types of library services using them. And when THE TECH SET won the ALA's Greenwood Publishing Group Award for the Best Book in Library Literature, it seemed that we had achieved our goal of becoming the go-to resource for libraries wanting hands-on technology primers. For these new ten books, I thought it was important to incorporate reader feedback by adding two new chapters that would better facilitate learning how to put these new technologies into practice in libraries. The new chapter called "Social Mechanics" discusses strategies for gaining buy-in and support from organizational stakeholders, and the additional "Developing Trends" chapter looks ahead to future directions of these technologies. These new chapters round out the books that discuss the entire life cycle of these tech initiatives, including everything from what it takes to plan, strategize, implement, market, and measure the success of these projects.

While each book covers the A–Zs of each technology being discussed, the hands-on "Implementation" chapters, chock-full of detailed project instructions, account for the largest portion of the books. These chapters start off with a basic "recipe" for how to effectively use the technology in a library and then build on that foundation to offer more and more advanced project ideas. Because these books are designed to appeal to readers of all levels of expertise, both the novice and advanced technologist will find something useful in these chapters, as the proposed projects and initiatives run the gamut from the basic how to create a Foursquare campaign for your library to how to build an iPhone application. Similarly, the new Drupal webmaster will benefit from the instructions for how to configure a basic library website, while the advanced web services librarian may be interested in the instructions for powering a dynamic library website in the cloud using Amazon's EC2 service.

I have had the pleasure of hearing Jason Clark speak at numerous library conferences and have always walked away inspired and impressed. I knew that if anyone in the field could offer wisdom and advice in the form of a practical guide to building mobile applications it would be Jason—and he didn't disappoint. Jason went above and beyond the parameters that I outlined for this title and actually built all of the applications discussed in the "Implementation" chapter and provided his code for readers to download and reuse. Librarians will walk away from this go-to resource having built complete mobile applications for their libraries.

Ellyssa Kroski
Manager of Information Systems
New York Law Institute
http://www.ellyssakroski.com/
http://oedb.org/blogs/ilibrarian/
ellyssakroski@yahoo.com

Ellyssa Kroski is the Manager of Information Systems at the New York Law Institute as well as a writer, educator, and international conference speaker. In 2011, she won the ALA's Greenwood Publishing Group Award for the Best Book in Library Literature for THE TECH SET, the ten-book technology series that she created and edited. She's also the author of *Web 2.0 for Librarians and Information Professionals*, a well-reviewed book on web technologies and libraries. She speaks at several conferences a year, mainly about new tech trends, digital strategy, and libraries. She is an adjunct faculty member at Pratt Institute and blogs at *iLibrarian*.

PREFACE

Mobile devices are becoming an essential part of people's everyday lives. As these new devices such as smartphones, tablets, and e-book readers move into the mainstream, there will be an expectation that library services and resources will be part of this mobile ecosystem.

Building Mobile Library Applications focuses on mobile application design and development—the practice of building software, web apps, or websites for mobile and handheld devices. Learning about mobile application development is one step librarians can take to answer the growing expectations for real-time, at-hand information consumption that mobile devices provide. Taking this a step further, mobile-savvy librarians are moving beyond just learning about mobile to actually building mobile library applications that provide patrons with catalog searches on the go, promote library databases optimized for mobile, and offer other cutting-edge services like historical walking tours using mobile devices. In learning to build and use these types of mobile applications, libraries can engage their patrons in context, in locations where they need the info.

ORGANIZATION

With both beginning and expert developers in mind, *Building Mobile Library Applications* guides you through the process of planning, developing, and launching your own mobile library applications. Chapter 1 traces the emergence of the mobile platform and introduces the possibilities for mobile development. Chapter 2 considers the types of mobile applications that can be developed and looks to guide development decisions by discussing what type of application makes sense for your mobile use case. Chapter 3 moves on to the details of project planning and processes that make sense for a mobile project work flow. Chapter 4 brings up the social aspect of mobile development and design, talking through how to garner support for your mobile project ideas and providing strategies for shepherding your mobile project through your organization. Chapter 5 focuses on the "how to" with a set of projects ready for implementation, including detailed code recipes and working downloads to get you started. These "takeaway" projects form the core of the book and provide an entry point to mobile development for all skill levels from beginner to expert. Included among the featured projects are learning how to develop an iPhone or Android application for your library, how to mobilize your library's catalog using a mobile web app, and how to create a mobile website that can be viewed on smartphones. Chapter 6 takes a closer look at how to market your mobile applications to your patrons, search engines, and mobile app stores or marketplaces. Chapter 7 considers emerging best practices and user interface conventions that make designing and developing for mobile an exciting challenge. Chapter 8 shows how to measure the success of a mobile app with analytics and statistical tools that tell the story of your app. Chapter 9 highlights the trends for mobile development and design in the months to come. Finally, the "Recommended Reading" chapter lists and annotates resources to continue learning about mobile design and development.

A primary goal of *Building Mobile Library Applications* is to demystify the process behind developing and designing for the mobile setting. And anybody looking to get a handle on what mobile means for libraries and related institutions will find this book to be a valuable guide. Learning about mobile technologies is a first step, and this book will cover the background of mobile devices, how to think about design for the mobile setting, planning for mobile projects, and much more. However, the core of *Building Mobile Library Applications* will focus on how to build sample mobile applications that use library data or work in a library setting. It is my hope that readers are empowered to create new library applications and services based on the code samples and walk-throughs available here.

ACKNOWLEDGMENTS

I would like to acknowledge the terrific work of Ellyssa, my editor. Her vision for the series helped frame my writing, her insistence on keeping the projects and writing accessible for a broad audience resulted in a better book, and her patience and consistent demeanor as I struggled with the creative process of writing was much appreciated.

▶ 1

INTRODUCTION

- Mobile Design and Development
- Who Should Read This Book? What Can You Expect?

With four billion applications on just one mobile device platform (iPhone) and device purchasing set to outpace both types of desktop computers combined, there can be little doubt that mobile is moving into the mainstream. Given this rapid adoption, my hope is that this book is a discussion of and a foundation for learning how to build mobile applications and sites.

Mobile Device Usage

- By the year 2014, consumers will be buying more smartphones than PCs and laptops.
- Since the launch of the iPhone, more than four billion apps have been downloaded, with an average of 47 apps per user. Android and iPad app stats are also in the millions.

From "Internet Trends," PowerPoint Presentation at Morgan Stanley's CM Summit, June 2010 (http://www.morganstanley.com/institutional/techresearch/pdfs/MS_Internet_Trends_060710.pdf).

Mobile applications, apps for short, are stand-alone, dedicated pieces of software or web applications/sites that enhance our phones' or tablets' capabilities and access information in elegant, consistent ways, and are the means for creating new services for our mobile patrons. People want apps; they have been trained to expect apps for their mobile devices. Library software development must keep up with the demand. We can gain much in this pursuit. Among the possibilities are:

- new ways of browsing using location data,
- real-time, contextual search providing results about where a person is located,
- voice-initiated browsing and searching, and
- archiving images and documents from mobile cameras.

In many ways, our success in reworking traditional library web services into mobile settings will help define the direction of our profession.

The rise of the mobile platform can be traced to Apple's release of the iPhone on June 19, 2007. With the release of the iPhone, consumers now had access to a mobile computer in their pocket. The smartphone template introduced by the iPhone changed what people expected to experience in the mobile setting. It wasn't just about texting or phone calls anymore; here was a computer with a full web browser and optimized operating system built for computing in mobile settings with limited bandwidth and connections. Portable media browsing, media creation (images and video), full website viewing, and other actions commonly associated with desktop PCs were now a part of the mobile environment. And apps, those little pieces of downloaded software or optimized web applications and sites, became the conduit for services delivered to this new platform.

Given the relative newness of the mobile platform, the history of mobile development in libraries is brief, but growing quickly as one might expect. One of the first libraries to enter mobile development was the District of Columbia Public Library (DCPL). In early 2009, DCPL built an app for browsing and searching library materials and released it for the iPhone (http://dclibrarylabs.org/archives/476). The DCPL app was a first attempt to translate a traditional library service, the catalog search, into a mobile setting. Three years later, the move to mobilize the catalog remains the most frequent mobile app type coming from libraries. A next step for libraries was to recognize the local context and immediacy of place that could be applied to mobile development. To this end, in early 2010 North Carolina State University (NCSU) Library released WolfWalk, an app based around a historical walking tour with archival photos of the NCSU campus (http://goo.gl/ga4YQ). As the mobile platform has matured, other cultural organizations have begun to experiment with mobile development. The Smithsonian Institution has a complete mobile development arm that is building apps ranging from Leafsnap, a mobile app that uses the device camera to help identify tree and plant species, to Stories from Main Street, a crowdsourcing mobile app that uses device microphones to record local history stories from all over the nation.

▶ MOBILE DESIGN AND DEVELOPMENT

Not all libraries will have the types of development resources mentioned, but each of us can get started with a basic understanding of the benefits and complexities of mobile design and development. First, mobile design and development can be liberating. Whitespace is necessary, and screen space is at a premium. Decisions about what to include in your mobile app or site need to be based on the core actions and utility your users need. This "limitation" of the small surfaces in mobile frees you from the complexity associated with the multiple links and entry points of desktop applications. Second, mobile design and development addresses an emerging need of our library audience: the ability to use library resources and get questions answered when the need arises. Mobile brings the dream of a portable library into reality. Third, mobile design and development can leverage existing skill sets. Many of the apps we build in this book will use HTML, CSS, and JavaScript skills that are already in place for many libraries. This "mobile web-centric" approach to mobile development offers a way forward that can make library resources truly cross-platform. Finally, mobile design and development and its simplicity aesthetic can inform physical library services. By forcing us to take a hard look at what is essential for a service to succeed, mobile can help us revise and reform current library services.

Even with these benefits, I'm not looking to trivialize mobile design and development. Creating simple mobile designs can be really difficult. Multiple devices and the growing fragmentation of the mobile market are huge design and development challenges. What works on one platform may not work on another. Additionally, having to choose a mobile platform—Apple (iOS), Android, BlackBerry—to provide library materials or to invest time learning a new software development environment can be cost-prohibitive or even run counter to the library mission of equal access for all. However, there are ways around these potential sticking points, and, whenever possible, I have looked to develop platform-neutral solutions for this book.

▶ WHO SHOULD READ THIS BOOK? WHAT CAN YOU EXPECT?

This book is for anybody looking to get a handle on what mobile means for libraries and related institutions. Readers should also have a keen interest in learning how to make decisions about a mobile strategy and getting their hands dirty with practical, applied mobile projects. At its core, this book is about the implementation of some exemplary mobile projects. These projects range from the simple to the complex, but all projects are written up in a tutorial, step-by-step manner. All you'll need to follow along with the vast majority of examples is a text editor and a web browser (recent versions of Internet Explorer, Chrome, Firefox, or Safari).

Over the course of this book, we will look at defining types of mobile apps, planning and project management for mobile development, negotiating the social mechanics of your library, marketing your apps and sites in this new and emerging mobile ecosystem, and discussing developing mobile trends. When we are finished, you will have a full sense of how to think broadly about mobile development and design. You will also have multiple working projects and examples of how to create mobile apps and websites for your library. Specific projects include:

- learning how to develop an iPhone application that features core library services,
- building a "Where's My Library" location-aware Android application using Google's App Inventor,
- mobilizing your library's catalog using WorldCat and its associated developer's tools, and
- creating a mobile website that can be viewed on smartphones.

There is something here for beginners and advanced developers, and the "cookbook" format will allow you to move from the simple to the complex. Let's get started.

▸2

TYPES OF SOLUTIONS AVAILABLE

- ▸ Investigate the Types of Mobile Applications
- ▸ Decide Which Type of Mobile Solution to Use

Mobile as a medium continues to evolve. The types of applications one can build into the emerging mobile platform follow this shifting trajectory. Even with this moving target, it can be useful to delineate the types of applications that you might consider as possible candidates for library mobile applications. This chapter will look at three primary types of mobile applications: mobile websites, mobile web applications, and native applications. We will define each mobile application type and weigh the pros and cons to help you decide how you might move forward with your mobile apps or services.

▸ INVESTIGATE THE TYPES OF MOBILE APPLICATIONS

Mobile Websites

When people begin to think about building mobile applications, a common first thought is to build a website that is optimized for mobile browsing. Mobile websites are websites specifically built to work within the mobile browsing environment. Typically, they have a simple list architecture, rely on text links and typography, and are built using simple HTML. Mobile websites were some of the first types of mobile applications to appear, as their simple text display worked in even the most primitive browsing environments, like first-generation mobile phones. From a content perspective, mobile websites tend to be informational in nature—a list of links that point to a secondary page of text—and lack any of the common interaction you might find in a desktop website. From a development perspective, mobile websites are fairly easy to create, as the foundation of the app is a common web language (HTML) that library staff or IT staff will have some familiarity and experience using from building traditional desktop applications. To see how you might build an example of a mobile website, see the "Mobilize Your Site with CSS" project in Chapter 5.

The pros of mobile websites include the following:

- They are simple to create, publish, and maintain.
- They make use of existing skill sets and knowledge that your library or organization has in place to build desktop websites.
- They use simple text and typography as the primary user interface, eliminating the need for complex design skills on staff.
- They create "lowest common denominator" views of your data that can be viewed on most mobile devices.
- Because they are "live" on the web, iterating designs and posting updates are quick and easy. There are no "app store" submission processes to slow down development.

The cons of mobile websites include these:

- Because they use the mobile web, connections and lack of network broadband can compromise performance.
- Because of their simplicity, they offer users a very limited experience.
- Many mobile websites are simple conversions of desktop sites without much consideration given to how the mobile user has different needs and expectations.
- Even with their simplicity, they can be difficult to support across all mobile devices.

Mobile Websites

- Chelmsford (MA) Public Library
  http://www.chelmsfordlibrary.org/mobile/
- Florida International University Medical Library
  http://medlib.fiu.edu/m/
- Iowa City Public Library
  http://m.icpl.org/m/

Mobile Web Applications

In response to the shifting nature of the mobile platform and the emergence of smartphones with advanced web browsers, developers have begun to create mobile web applications that bring application-like experiences into the mobile browser environment. Mobile web applications use HTML, CSS, and JavaScript as well as the mobile web network and web browser to give a native application feel to mobile websites. By "native application feel," I mean mobile web applications that create a user experience that mimics the way that an installed smartphone app behaves. Moving away from the simple typography and textual content page views of mobile websites, mobile web apps allow the user to browse content in real time within a continuous page view or "frame" that is updated as a user touches or pinches the device. With the reliance on the mobile web as the carrier for their data, these applications do not need to be downloaded or compiled on a target device. To see how you might build an example of a mobile web application, see the "Mobilize Your Library's Catalog" project in Chapter 5.

The pros of mobile web applications include the following:

- Because they are "live" on the web, content is accessible on any device with a mobile browser.
- They are fairly simple to create and maintain using basic HTML, CSS, and JavaScript.
- Because they are not tied to a specific platform, debugging, redesign, and updates are simple. There are no "app store" submission or review processes to slow down development.
- With advanced functionality and behaviors (usually supplied by JavaScript), they offer a richer experience more in line with mobile user expectations.

The cons of mobile web applications include these:

- Because they use the mobile web, connections and lack of network broadband can compromise performance.
- Content is accessible on any device with a mobile browser, but ensuring that performance or behavior is consistent across all devices can be tricky.
- They don't always have access to native application functions like geolocation, camera use, offline storage, local device files, and so forth. (This is starting to change with HTML5.)

Mobile Web Apps

- Oregon State University, BeaverTracks (historical walking tour)
  http://tour.library.oregonstate.edu/
- University of Minnesota Library
  http://www.lib.umn.edu/mobile/
- McGill University Libraries
  http://m.library.mcgill.ca/

Native Applications

The third type of mobile application is the native application. When people speak of "apps" they are likely referring to native applications, as these apps were some of the first introduced with native calendaring and e-mail applications on specific devices. Native applications are installed and run on a device and are developed using the software development kit (SDK) for a specific platform (e.g., Google's Android or Apple's iOS). Native apps offer a rich user experience, because their programs are able to tie into local device hardware and software functions like the camera, messaging/notification capabilities, geo-location tracking, offline modes, and local file system access. With this reliance on a specific platform and SDK, native apps can run only on a single targeted device. This means you have to develop a different application for every platform. However, the reliance on a single platform and SDK can also have benefits. Native apps are often regulated and vetted for quality after a submission and certification process, lending a professional and polished nature to the performance and behavior of the application. Once a native app is built, it is usually distributed or sold in its respective marketplace or app store. To see how you might build an example of a native application, see the "Build an Android App" or "Build an iPhone App" projects in Chapter 5.

The pros of native applications include the following:

- With access to local device hardware and software, they offer the most seamless and richest user experience for mobile applications.
- They use proprietary distribution models or channels that allow you to charge people for your application.
- With the help of software frameworks like PhoneGap, Sencha Touch, and Titanium, native applications are becoming easier to build.
- The certification and submission processes associated with native applications ensure high quality and security.

The cons of native applications include these:

- Because they are tied to a specific platform and SDK, debugging, redesign, and updates can be complex and onerous.
- Developing across multiple platforms and learning specific programming languages for each targeted mobile device is costly and requires significant resources.
- The certification and submission processes for native apps can be complex and not fully documented.

- While you may charge people for your native application, you may need to share revenue with the third party providing the marketplace or app store.

DECIDE WHICH TYPE OF MOBILE SOLUTION TO USE

Native Apps

- ugl4eva
  Video tour app of the Undergraduate Library, University of Illinois, Urbana-Champaign
  http://itunes.apple.com/us/app/ugl4eva/id352224134?mt=8
- Seattle Public Library
  http://itunes.apple.com/us/app/spl-mobile/id364019201?mt=8
- BookMinder Android App
  http://www.oclc.org/developer/prototypes/bookminder

Your choice for the type of mobile solution you put in place for your library or organization will depend on your resources (human, fiscal, and knowledge) and your users' needs. The mobile application types defined earlier in this chapter will get you started thinking about the types of solutions available and why you might decide on one approach over the other. Here, we'll look at the particular uses cases for mobile projects so that you can choose wisely. Keep in mind that you may need to use more than one mobile application type to reach most of your patrons.

Many discussions regarding what type of mobile projects to build focus on the question of native apps versus web apps. As is usually the case, it's never black and white, and this is a bit of a false dichotomy. When we place the question in the context of library applications, things become even murkier. For example, questions of universal access to information, which form core missions for many libraries, could restrict choices to support a single mobile platform with your app. Do you support Android or iOS (Apple)? Fortunately, there are a few methods that can guide library developers as we start to pick and choose the types of mobile projects we might build.

One method for selecting the type of mobile project is to try to answer the question: "What do my users need to do?" This is a very common approach, as it lets the developer sketch out a persona, an imagined, prospective user, and let that persona guide what type of application you might build. Here's a quick narrative flowchart for project selection based on how a user needs to interact with your library application.

- People need static info about the library—hours, contact info, location → Build a mobile website.
- People need to search the catalog, link to mobile databases → Build a mobile web app.
- People need to be able to use device camera and device orientation to navigate and contribute to a local history app → Build a native app.

Another method to help you in your decision is to sketch out the application's requirements and let those requirements guide your development choices. App requirements might typically include the mobile platform one must support, the technical expertise needed to build the app, and types of data needed to make the app run. With this emphasis on app requirements rather than on the actual users of your app, your decision tree could look something like the following:

- My app needs to work across platforms and must be able to use HTML (based on in-house expertise) → Build a mobile website.
- My app needs to work across platforms, load dynamic data from our digital collections, and use geolocation to place geotags on these records → Build a mobile web app.
- My app needs to work without an Internet connection, be able to be monetized, use local device hardware (camera, phone, etc.), and be distributed in the app store or app marketplace → Build a native app.

It is important to note that the decisions we tracked here are primarily suggestions. Chances are you will want to use a combination of user and app requirements as well as cues from library administration to decide which type of app (or site) to build that works best for your organization and patrons.

3

PLANNING

- Define Your Audience and Research Your Mobile Market
- Define Your Primary Reason for the Mobile Project
- Define Your Primary Goal for the Mobile Project
- Outline Mobile Project Requirements and Deliverables
- Define Your Working Group or Team and Assign Roles
- Set Up Milestones and Timelines for Your Mobile Project

While mobile projects can vary greatly, there are some basic project planning steps and techniques that can be used to help a project move toward a successful conclusion. Over the course of this chapter, we will take a closer look at these steps and techniques. Our goal is to show some basic mobile project planning guidelines that will help you think more broadly about your mobile apps and services and create pathways for a successful project.

As we move forward, I will be referring to separate steps in the process, but I would encourage you to collect what you learn and find in each step a single project definition document. If you haven't already, open up an MS Word or a Google Docs document to collect your definitions, timetables, and findings. Let's get started.

DEFINE YOUR AUDIENCE AND RESEARCH YOUR MOBILE MARKET

A successful project begins with an understanding of the needs of your primary audience and knowing the demands of your specific market. Your first step is to define your primary audience. Ask yourself or your working group: "Who are the users of the mobile app or service?" Your answers will inform the direction of your project and your research. For example, if you identify a library researcher as primary, you will come up with a different set of app objectives and research questions than if you identify young adults as your primary audience. Try to limit it to primary users in your initial answers.

Once you have decided on a primary audience, you can move on to ask questions and do some research about the needs of that audience. Research can take many forms: informal conversations with patrons, open surveys, formal usability testing, web statistical analysis, and so forth. We are going to focus on two methods that you can implement quickly: a survey and a web stats analysis. With each of these methods, we will be looking for answers to the following questions:

1. What mobile devices are your patrons using?
2. What is the top content on your library website?
3. Can the top content be translated/transformed for the mobile setting?
4. What library content do your patrons expect to use in the mobile setting?

A survey using an online survey generator, like SurveyMonkey (http://www.surveymonkey.com/), can be a very effective way to gather the opinions of library users. Using SurveyMonkey, you can construct a public survey with variations on all of the listed questions. You can also use Google Docs Forms (https://docs.google.com/support/bin/answer.py?answer=87809) to create a series of questions and collect the data. Promote the survey on your website or in other public forums. If you have the opportunity, try to ask questions of your primary audience directly. For instance, e-mail active library researchers with the survey link and a quick explanation of your research to get direct insight into primary audience needs. Leave the survey open until you feel you have sufficient data to make a decision. When finished, build a report using the data, and move forward to the next step in the planning process.

Web statistics and server logs are data sources that are available from most web servers and are an invaluable resource in making decisions about your mobile app or service. You can mine these collected stats to help answer many of the questions listed. Several metrics included in your server statistical logs will help:

1. **User agents** will answer the question of what type of devices your patrons are using.
2. **Top page views** will show you the most visited content.
3. **Page size and loading times** can help identify content that will convert well into the mobile setting.

Most web statistics packages record these metrics. If you don't have access to your web statistics, ask your server administrator for help in generating a statistics report. Hosted options, such as Google Analytics (http://www.google.com/analytics/), do have these metrics in place and provide a simple interface for you to build reports. The key is to decide on your preferred tool for gathering statistics and making sure it is installed and collecting data.

Your research will be foundational in how you represent and make the case to others in your library for creating an app. Data results from the methods described will ground your argument and allow you to move away from the anecdotal when you talk about the reasoning behind your mobile app or service.

▶ DEFINE YOUR PRIMARY REASON FOR THE MOBILE PROJECT

Being able to articulate "why" your mobile project makes sense as an investment is another essential piece of project planning. An analysis based on your research or on mobile trends in general can help justify pursuing a mobile app or service. The project leader or manager needs to be able to speak to all library staff to help them understand why this project is worthwhile. To this end, use your research and recognize larger mobile trends in crafting a thesis statement (of sorts) for your mobile app or service.

Sample Thesis Statement

The mobile platform represents a fundamental shift in expectations and a new form of mass media. In our research, we are seeing a clear preference for library information and the need for access to library resources in local and immediate contexts. In an effort to serve our patrons, we are building a mobile app that will answer these emerging user needs.

Particular wording will depend on what makes sense for your organization and your specific project, but drafting a reasoned statement explaining why your mobile app or service is an important investment will help define the project aims and goals.

▶ DEFINE YOUR PRIMARY GOAL FOR THE MOBILE PROJECT

Knowing the scope of your project and composing a description that defines and limits the primary actions of your mobile app or service is essential. Scope creep, or the tendency for additional functionality to be added to the project, can create bloated apps and services. Think of your favorite mobile apps or services. Typically, they do one or two things really well. Apply this lesson to your own development. Define a singular goal for your project initially. In our case, it might be something like: "Our app will aid in discovery of library resources for mobile settings. Included in our discovery layer will be the ability to search the catalog and renew items." A primary action or two is all you need at the outset of the project.

▶ OUTLINE MOBILE PROJECT REQUIREMENTS AND DELIVERABLES

Based on your research and project goals, you must decide what final result or product you will create with your mobile project. It is best to limit the deliverable to a simple product in any of your initial sketches. A deliverable is the quantifiable result of all of the project work. In a library mobile project, it might be the app or service itself. It is useful to outline key requirements and details as you work toward sketching what the project deliverable will be. Some example requirements include the following:

- Technical specifications: programming languages, mobile platform supported (Android, iOS)
- Branding requirements
- Maintenance and update routines
- Project meeting schedules during development

Keep it simple, as complexity will come over the course of development, conversations with stakeholders, primary audience feedback, and so forth. When you have your key requirements in place, you can refine your project deliverable and specify what the end result of the project work will be. For example, the deliverable statement that might appear after the list of requirements could read: "Deliverable: an Android application that uses JavaScript to search the library catalog using library branding and has quarterly development updates."

▶ DEFINE YOUR WORKING GROUP OR TEAM AND ASSIGN ROLES

A mobile project can work in a team or individual developer setting. Defining the staff possibilities for your mobile project will depend on your organizational resources. Clearly defining the roles that are needed for the project and deciding who will take on those roles is the first step. A typical mobile app or website project will have the following roles:

1. Designer
2. Developer
3. Copywriter
4. Project Manager

You may have to take on all the roles depending on your staff resources. Decide on the roles for your project, and list them in your project definition document.

▶ SET UP MILESTONES AND TIMELINES FOR YOUR MOBILE PROJECT

A successful project is a timely one. Estimating the work needed to complete a project and setting dates when the work will be finished is one way to keep your mobile project on track. A typical milestone template or draft might look like this:

Milestone 1 (2 months): User Research—Date Complete: 7/01/2011
Milestone 2 (3 months): Application Development—Date Complete: 10/01/2011
Milestone 3 (1 month): Application Testing—Date Complete: 11/01/2011
Milestone 4 (1 month): Rollout and Promotion—Date Complete: 12/01/2011

The idea is to set a task, prescribe a time limit, and set a due date. (Specifics will vary depending on the details of your project.) This is not to say that there won't be sidetracks or bugs that will slow you down; most projects are rarely linear. But, estimating and defining tasks and times will set expectations for the project and allow you to communicate with your team (or yourself) and your organization what to expect from their investment.

Mobile Design and Development Planning Documents

Two resources have been particularly useful in refining my thinking on planning for mobile:

1. Luke Wroblewski's thoughts on project definition with sample templates at http://www.lukew.com/services/project_definition.html
2. The Project Planning Templates from Usability.gov at http://www.usability.gov/templates/index.html

---

▸4

SOCIAL MECHANICS

- Articulate and Define Your Project Goals and Deliverables
- Tailor Your Discussion to Each Group of Stakeholders
- Create a Supporting Network

Any discussion about mobile apps and services will need to focus on the project details, but there are many other considerations, most notably, the social considerations of building support in your library for the project. In this chapter, we will take a look at ways to build the case for your mobile apps and services and to help create social infrastructures and networks that support your project.

▸ARTICULATE AND DEFINE YOUR PROJECT GOALS AND DELIVERABLES

Recognition of your role as an advocate for and an expert about the mobile app or service you want to put in place is an essential move. Every idea or project needs a champion, and your first role will be considering what it means to articulate the reasons behind the project to all the organizational levels of your library. Thinking in broad terms here is useful. These questions will get you started:

- How would you explain your mobile app or service idea in 160 characters or less?
- What technologies are needed to build and support your mobile app or service?
- How does your mobile app or service support current organizational priorities?
- Why are other peer institutions building and supporting similar mobile apps or services?

- How do your web statistics or other analytics demonstrate public interest in your proposed mobile app or service? If they don't, how else might you justify your proposed project?
- What is your project deliverable, timeline for implementation, and plan for promotion of your mobile app or service?

It comes down to doing your research. You need to be able to speak intelligently about the scope and shape of your project as well as the reasoning behind your proposal.

Once you have done enough research to take on the role of advocate or expert, you will want to look to generate interest and buy-in from stakeholders. All successful projects need allies. Think long and hard about the individuals or teams you might need to understand and support your proposed mobile app or service. In a library setting, potential allies could be library administration, librarians, library staff, systems administrators, and your library patrons themselves. One method of presenting your idea to potential stakeholders is to prototype your mobile app or service. A prototype is a rough sketch or draft of your app or service. This visual demonstration of your app could be very simple or more elaborate if you have time and resources. On the simpler side, you could create a rough sketch of your app in a notebook with each page representing a different view or action for the app. Low tech can work effectively. On the more complex side, you could create an HTML prototype that shows how your app will function in a more realistic setting. In either case, the idea is to provide a demo that gives your audience a visual sense of what your app or service will do. Once the vision of your app or service is in place, it becomes easier to explain and to gauge the interest of potential stakeholders. When showing your prototype, look for an excitement about the project and a keen understanding of what your app can do from these potential allies. Invite further participation from those individuals who show this excitement and intuitive sense about your project.

Beyond the visual prototype idea, communicating with primary stakeholders on their terms is another means of building support for your proposed mobile app or service. In this instance, it is helpful to think about how to frame your discussion based on the unique questions that each stakeholder might have about your mobile app or service. Let's take a closer look at those potential stakeholders we identified earlier to see how their perspective can inform how to build the case for a mobile app or service.

▶ TAILOR YOUR DISCUSSION TO EACH GROUP OF STAKEHOLDERS

Library Administration

Your direct manager and upper levels of library administration will have a macro view of the library and will carry this view into the consideration of your mobile app or service. When you have the opportunity to talk with these stakeholders, you should appear confident, knowledgeable, and articulate about the objectives and plan for your project. Model yourself as a project leader throughout this exchange. Possible questions might include these:

- Why should we put resources in mobile development?
- Are peer institutions putting similar services in place?
- How will your project improve existing services and library functions?

Librarians and Library Staff

In most cases, these stakeholders will be your direct peers. They will have a grounded, somewhat anecdotal, perspective of how patrons are using library services and technology. This firsthand knowledge can be very valuable when thinking through design and development of a mobile app or service. When addressing these stakeholders, you need to be able to explain your app concisely and invite them to help shape the scope and direction of the project. The primary work of your project will be completed with the help of this group, and it is in your interest to enable their investment in your idea. Allow them to participate in the creative processes associated with your mobile app or service. Possible questions might include these:

- What roles can they play in shaping this new mobile app or service?
- How does it improve the work flow of library users?

Systems and Library IT

Your library systems team and systems administrators are another essential constituency. In the context of mobile apps or services, these stakeholders will be some of the most knowledgeable. They will also be the linchpin in any mobile technology program you might propose as they will be maintaining and supporting the systems with which you will build your app. In working through your idea with this group, you will want to appear confident and show that you have done your homework. Demonstrate your expertise. Possible questions might include these:

- What technologies do you need? Are they part of your current systems environment?
- How do you see your app or service scaling up with increased use?
- What are your plans for maintenance and continued development of your mobile app or service?

Library Patrons

Your library patrons are less visible but are a very important silent majority. Find ways to incorporate their needs and feedback into your mobile app or service. Surveys, focus groups, usability tests, and the like—all are possible means to addressing these stakeholders, but that will depend on your time and resources. When working with this group, you will again need to establish your expertise and clearly explain how their commentary will directly shape the development of your app or service. Possible questions might include these:

- How does your app answer a need a patron may have for a new or improved library service?
- Does your app save patrons' time?

▶ CREATE A SUPPORTING NETWORK

Being able to anticipate alternate perspectives about your mobile app or service and answer the questions posed in the previous section gives you a sense of how to best address the concerns and interests of your organization. It is important to remember that with a new mobile app or service you are introducing a new technology into your library. And in order to support a new technology, all levels of the library need to be able to understand its impact on their position and how it fits into the larger goals of the library mission and core services.

Until now, we have been focusing on building a case for mobile, but there is another equally pressing social mechanics need: creating social infrastructures and networks that can help support your mobile app or service. Two strategies that can help in this regard are setting up a mobile interest/working group and creating visibility for your mobile projects using promotions and soliciting feedback.

With your core stakeholders selected, it is time to establish a working group or an interest group related to your mobile app or service. In smaller libraries, your working group may consist of you and one other peer. Either way, you will want to think about establishing a membership, setting a meeting schedule, and creating a project management routine. During active development of your app or service, you will want to meet more frequently to keep things moving (e.g., once a week). As your app or service is released, you may want to scale back and meet on a monthly basis. In the end, you want to create a friendly social forum where a project can be managed and new ideas for mobile can be vetted and considered.

Finally, look for ways to promote and market your app or service to your remotely invested stakeholders. Marketing in this case is not about an active campaign. It is about keeping your app or service visible to those outside of your working group or peer advocates. If you can raise awareness of your mobile efforts, you can help to create advocacy moments in unlikely places. Potentially, your mobile message can be amplified by anyone in the library. See the following for some effective methods for raising awareness and creating a mobile advocacy culture in your library.

Creating Awareness about and Participation in Your Mobile Projects

- Announce progress or news of your mobile app or service at general meetings with all library staff.
- Schedule a brown-bag discussion (open to all library staff) related to mobile trends and developments.
- Blog about your app or service. Add a post about your mobile efforts to your library social networking site. If you have a library newsletter, submit a blurb about your app or service.
- Invite library staff or administration into a brainstorming session about possible ideas for your next mobile app or service.
- Set up a comments form on the public website and on your staff intranet soliciting feedback about your mobile app or service. Additionally, you could ask for new app ideas with the same form.
- Using discretion, e-mail useful links or writings about mobile trends and development. Start a collection of mobile links on the staff intranet that anyone can browse.

5

IMPLEMENTATION

- Create a Basic Mobile Website
- Build an Android App
- Build a Mobile Website from Scratch
- Build an iPhone App
- Mobilize Your Library's Catalog
- Mobilize Your Site with CSS
- Build a Mobile Site Using JavaScript Frameworks (jQuery Mobile)

The seven projects in this implementation chapter are organized around an "easy to difficult" theme. The idea is to allow readers to work through the sequence and build skills and more complex projects. Of course, this sequence is not required, and each app "recipe" can be implemented on its own. In addition, each of the major types of mobile applications discussed in Chapter 2—websites, web apps, and native apps—are included.

CREATE A BASIC MOBILE WEBSITE

Mobile site and app generators offer everyone the opportunity to create a mobile view of their library data. In this project, we will use the Winksite mobile site generator tool to create a mobile site using an RSS feed from a WordPress or Drupal content management system (CMS). If your library doesn't use either of these systems, you can grab any RSS feed to practice creating this project. Our finished project will look something like Figure 5.1. Let's get started.

▶ Figure 5.1: Example Mobile Site Generated Using Winksite

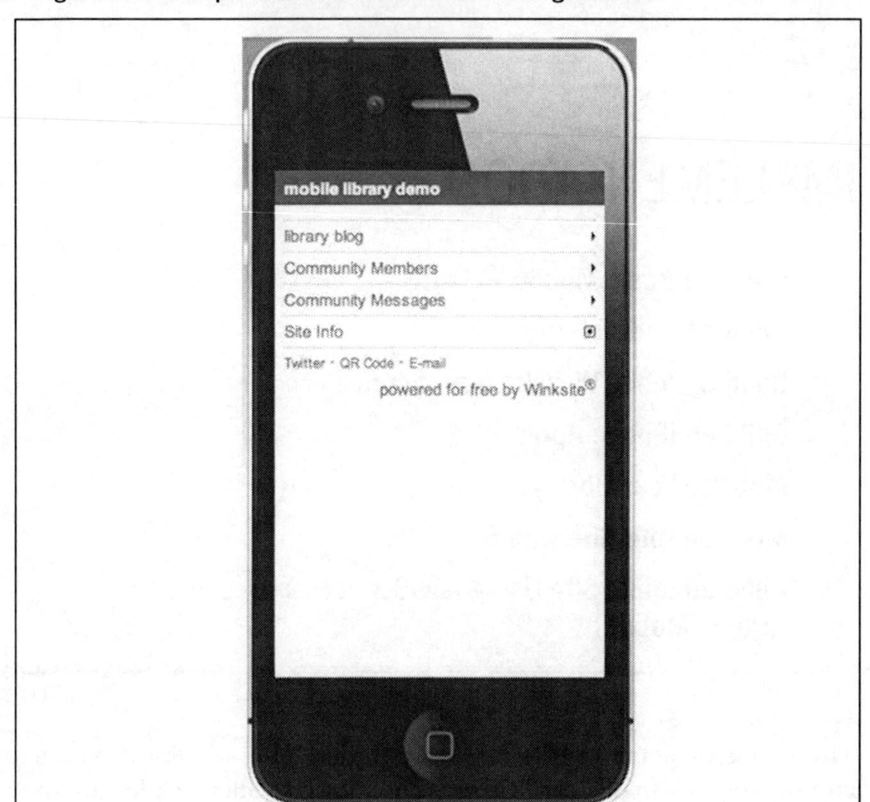

Step 1: Set Up a Winksite Account

We will need to set up an account with Winksite to use the mobile site generator tool. Winksite is free, and each account you set up using the tool is allowed to have up to five distinct mobile sites. Navigate in your web browser to the Winksite sign-up page at http://winksite.com/site/signin.cfm?action=signup. Fill in the registration form to set up your free Winksite account. One important note to keep in mind is that your Winksite account username will be part of your URL. I would suggest using a keyword from your library name as your username rather than a personal name.

Step 2: Find the Source URL for Your RSS Feed on Drupal or WordPress

For this project we are going to "mobilize" one of our library site's RSS feeds. Many content management systems and library site building software have feeds associated with the content created for each page within the site. We are focusing on two of the most popular CMSs in libraries, Drupal (http://drupal.org/) and WordPress (http://wordpress.org/), for this project, but any RSS feed will work as a source in this case.

Each Drupal site has a front page RSS feed that can be found at http://yoursite.com/rss.xml. Every WordPress site has a front page RSS feed that can be found at http://yoursite.com/?feed=rss2. WordPress has great documentation, and more information about the types of feeds you might create using WordPress is available at http://codex.wordpress.org/WordPress_Feeds. Use the appropriate URL structure provided to navigate to your CMS's RSS feed URL and copy it to your clipboard.

Step 3: Use the Winksite Dashboard Interface

With your RSS feed source URL on the clipboard, you are now ready to start using Winksite's dashboard and form wizards to create your mobile website. Using the Winksite account you created in Step 1, log in to the Winksite dashboard. Once you are logged in, you will see a button labeled "Create Site" in the middle of the page. Click on the button to get started setting up the parameters for your mobile site. You should see a screen like the one in Figure 5.2.

Once you have the dashboard view, it is simply a matter of filling in the required form fields to create a mobile site using your RSS feed. Add your site title, choose a unique and short site address, and provide a description for your site. Next, look for the "Mobilize Your Existing Blog" field. This is the field we will use to mobilize our RSS feed. Check the "Blog" checkbox to turn on this option. Set a title for the RSS feed, and, finally, paste your RSS feed URL in the "RSS Feed URL" text box. The Winksite dashboard has many options for creating different mobile page views. We are going to pass by these options right now, but you can always return to edit and add new pages to this site later. Select a category from the "Category" dropdown field, and then click on the "Build Site" button. You have just created your first mobile site using Winksite.

Step 4: Customize Your Mobile Site Using the Winksite Customization Wizard

After you have submitted your baseline information for the site, you will be greeted by the Winksite "Personalize Your Mobile Site" page,

► **Figure 5.2: View of Winksite Dashboard**

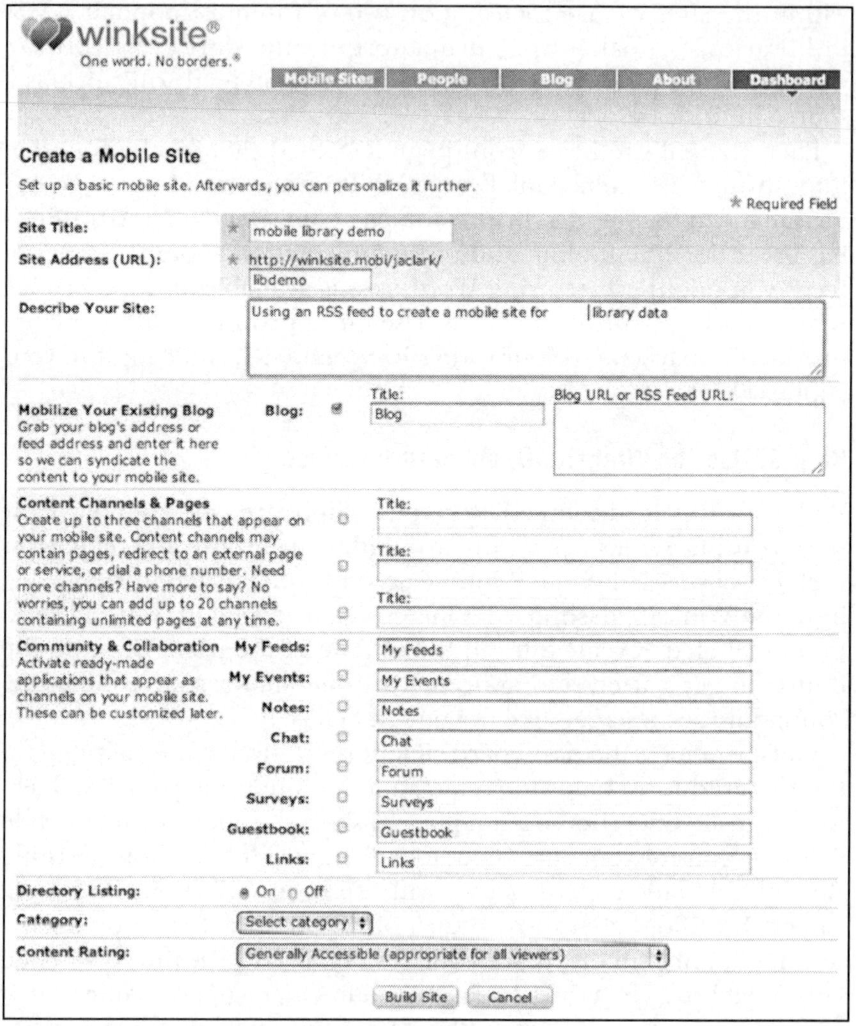

winksite®
One world. No borders.®
Mobile Sites | People | Blog | About | Dashboard

Create a Mobile Site

Set up a basic mobile site. Afterwards, you can personalize it further.

* Required Field

Site Title: * mobile library demo
Site Address (URL): * http://winksite.mobi/jaclark/ libdemo
Describe Your Site: Using an RSS feed to create a mobile site for library data

Mobilize Your Existing Blog Grab your blog's address or feed address and enter it here so we can syndicate the content to your mobile site. **Blog:** Title: Blog — Blog URL or RSS Feed URL:

Content Channels & Pages Create up to three channels that appear on your mobile site. Content channels may contain pages, redirect to an external page or service, or dial a phone number. Need more channels? Have more to say? No worries, you can add up to 20 channels containing unlimited pages at any time. Title: — Title: — Title:

Community & Collaboration Activate ready-made applications that appear as channels on your mobile site. These can be customized later.
My Feeds: My Feeds
My Events: My Events
Notes: Notes
Chat: Chat
Forum: Forum
Surveys: Surveys
Guestbook: Guestbook
Links: Links

Directory Listing: On Off
Category: Select category
Content Rating: Generally Accessible (appropriate for all viewers)

Build Site | Cancel

which will allow you to customize the look and feel of your mobile site. The page will look like Figure 5.3.

Make your way through the options. You could include a library logo if you have one. You can adjust the header colors to resemble your desktop library website. You can assign a "Link Method," which will designate whether to create links that link to internal pages within the mobile website or link out to external pages on the web. You can even upload a background image to replace the simple white page background that is set as the default. All of these customizations

▶ Figure 5.3: Customization Screen from Winksite

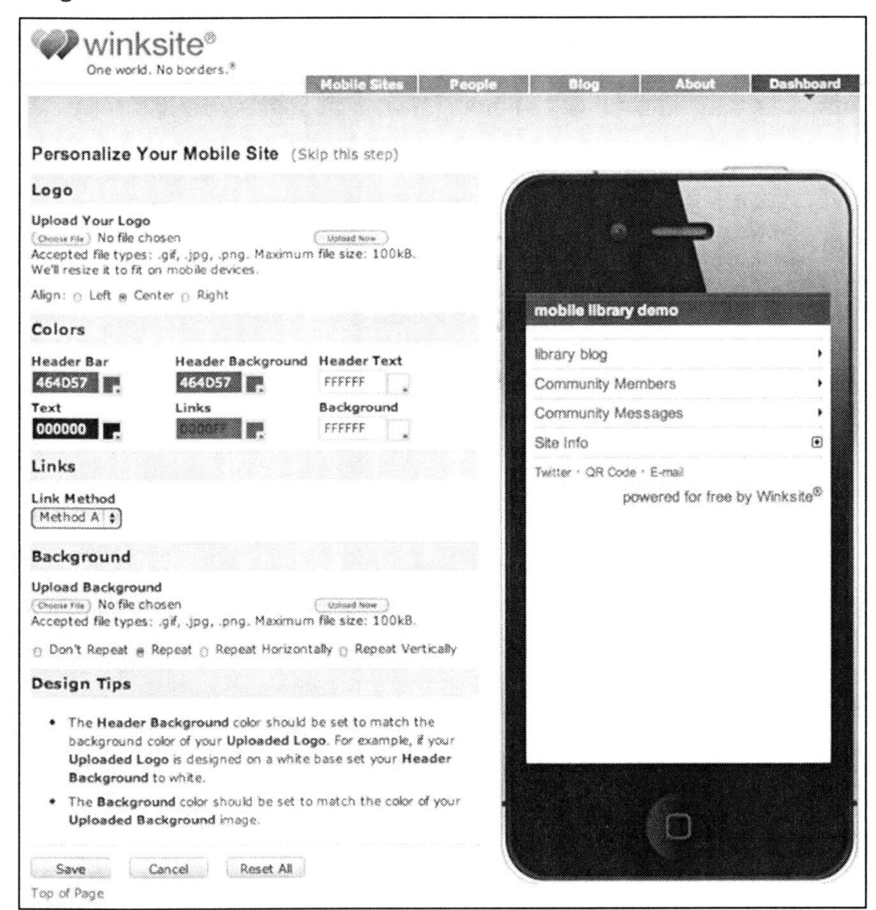

are optional, so feel free to choose only the ones that make sense for your design. When you are finished, click the "Save" button.

Step 5: Get Your Mobile Site URL, and Promote the URL to Your Patrons

After you have saved your mobile site, you will be given a view of your finished site and the URL for you to use as the public URL for your patrons. The view is the same view we looked at in Figure 5.1. Typically the URL convention will be: http://winksite.mobi/YOUR-USERNAME/YOUR-SITE-ADDRESS. This confirmation page also includes a QR code that you can use to direct people to your mobile site (see Figure 5.4).

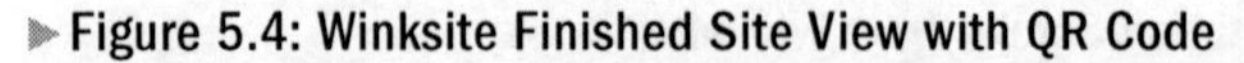

► Figure 5.4: Winksite Finished Site View with QR Code

And with that, you have created your first mobile website. Once you have tweaked the settings and are satisfied with the design, start promoting your mobile site on the front page of your main library website. You might also include a link on every page of the main website by placing a link to the mobile site in its header or footer. Promotion and marketing your site could include a quick blog post explaining what people can expect to find on your mobile site.

► BUILD AN ANDROID APP

View App Inventor here: http://info.appinventor.mit.edu/

Download the App Inventor "Where's My Library" code here: http://www.alatechsource.org/techset/; filename = wheresMyLibrary.zip (This code needs to be imported into your own App Inventor account to work.)

More and more tools that offer WYSIWYG text editors and graphical interfaces are being created to encourage people from all developer skill levels to become app developers. Using these tools can make application development feel more familiar, almost like using Adobe Dreamweaver. Google introduced "App Inventor" (http://info.app inventor.mit.edu/) exactly for this reason. (Note: Beginning in January 2012, the MIT Center for Mobile Learning will begin maintaining and developing the Android App Inventor. The functionality of the software will remain the same, but the look and feel and some URLs may change over time. Full details on the transition can be found at http://goo.gl/Mmn03.)

For this project, we are going to use the Android App Inventor to modify some existing source code to create a library application specifically designed for Android phones. The original app allows users to mark where they are on a map and get directions back to that spot from wherever they are currently located. We are going to modify the code slightly to mark where your library is in relation to a person's current location. When finished, our app will provide users with directions to the library from wherever they are currently located. This will give us an introduction to Google's App Inventor and teach us how we might use it to build more advanced applications going forward. Let's get started.

Step 1: Prepare Your Local Computer for App Inventor

Log in to App Inventor (http://info.appinventor.mit.edu/) using a Google account. Follow the steps to prepare your local computer for using App Inventor. You will also need to download and set up the App Inventor Setup Software. Find the link to the App Inventor Setup Software for your operating system at http://info.appinventor .mit.edu/. Most of the preparation will be testing for Java functionality and defining the emulator or device with which to test your app. An emulator is a "practice" environment that you can run as software on your local computer to see how your application will work when it is live on a mobile device. Choosing the emulator is a good idea, as it means you can test without having to have an Android device connected or on hand. App Inventor's emulator mode is unique in that it also allows us to use the emulator for building or putting together the components in our project. Go ahead and select the second option from the second step, "Build Your First App with the Emulator."

Step 2: Get the Android, Where's My Car Source Code

To get a feel for how App Inventor works, we are going to customize some existing app source code provided by MIT in their App Inventor tutorials available at http://info.appinventor.mit.edu/. As you get more comfortable with App Inventor, you might look through some of the other available tutorial projects for more ideas about what is possible to build. The app we are going to use is "Android, Where's My Car?" Download the source to your local computer from http://appinventor.googlelabs.com/learn/tutorials/whereismycar/AndroidWhereAssets/wheresMyCar.zip. (Note: These links to .zip tutorial source files have changed and will be linked from http://info.appinventor.mit.edu/.) Once you have downloaded the .zip file, upload the .zip file to App Inventor by clicking on the "My Projects" link and then choosing the "Upload Source" option under the "More Actions" dropdown.

Step 3: Learn the App Inventor Designer View

You should now have a "WheresMyCar" project listed in your App Inventor interface. Click on the project name to open it up in the App Inventor Designer view. The Designer view is the graphical user interface we will use to customize the app source code and create our Where's My Library app.

There are four columns in the Designer view (see Figure 5.5). Each column controls a different piece of the application:

- **Palette:** Includes ready-made options like input buttons, animations, and sensors for interaction events
- **Viewer:** Displays what these options will look like on the device screen
- **Components:** Provides a hierarchical listing of the assigned interaction events
- **Properties:** Holds the list of top-level information about the app, such as the primary title, app icon image, and orientation

Step 4: Customize the Android, Where's My Car Source Code

First, we are going to make a copy of the WheresMyCar project by using the "Save As" option. Rename the new project "WheresMyLibrary." Open up your new WheresMyLibrary project in the Designer view. Let's customize the name of our app by changing the title value to

▶ **Figure 5.5: Screenshot of the App Inventor Designer View**

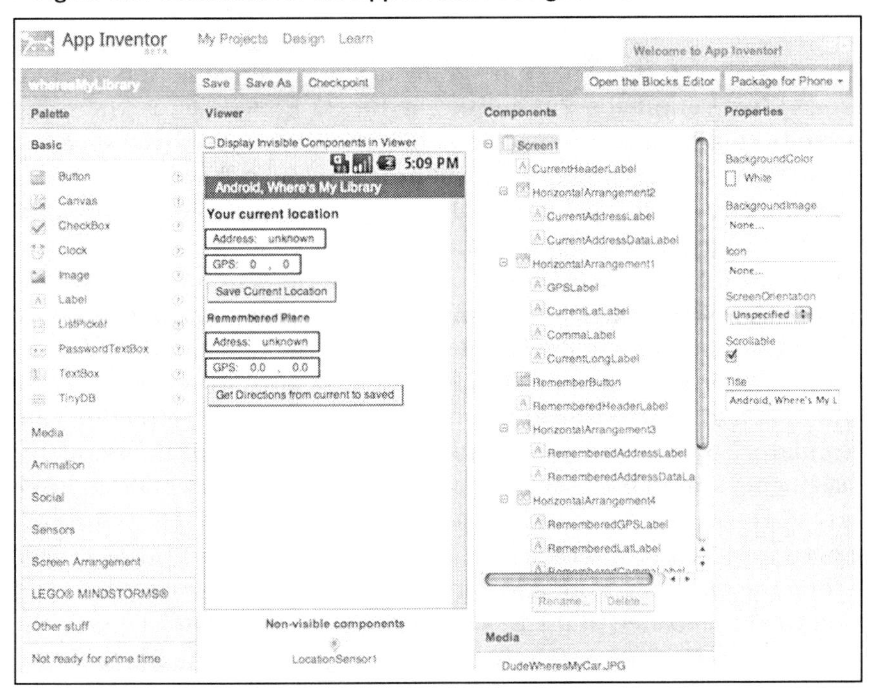

"Android, Where's YOUR-LIBRARY-NAME" in the Properties column on the far left. Enter your library name in place of YOUR-LIBRARY-NAME here. This label will be the title of the application in Google's App Market and the title that your users will see when they load the first page of your app. Next, let's change the input button text. An input button is a user interface utility common to many software interfaces that allows the user to send a command to a software program by clicking or touching a simulated button. If you have ever clicked a submit button on an HTML webform, you have used an input button. Click on the first input button in the App Inventor Viewer column named "Remember Current Location," and you will see the Properties column change. Find the "Text" field and change the label to "Save Current Location." Follow the same routine to change the second input button label to "Get Directions from Current to Saved." When touched (or interacted with by a user), this second input button records and stores your current location. Once both current and saved locations have been stored by the app, directions can be calculated and presented to the user. Finally, save your project.

Step 5: Test the Android, Where's My Library App with the Android Emulator

Once the project is saved, we are going to run it within the App Inventor Code Blocks emulator. Put simply, the Blocks Editor view is where you assemble program blocks that direct how the components (you specified while using the Designer view) will function and behave. You work to assemble programs in a visual setting, connecting components together like pieces of a puzzle. Look for the "Open the Blocks Editor" option in the top navigation bar and click on it to initiate a Blocks Editor download. Once the download is complete, a new browser should appear with the Blocks Editor view. In the top navigation for the Blocks Editor, you will see a "New Emulator" option. We are going to use this option to test the view and performance of our app. Select the "New Emulator" option and you should see a prompt about the emulator starting up and to hold on as systems are prepared. Click "Ok" to move past the message. Once the emulator is loaded, you will see an Android phone prototype. There is one final step to see your app in action. When the emulator is up and running, you need to connect your app that you have loaded in the Block Editor into the emulator. Choose the "Connect to Device" option in the top navigation of the Block Editor view, and pick the emulator you are running. (It will usually have a generic name like "emulator-5554.") Once connected, this emulator will be able to show you how your app will look and perform in a basic Android phone setting. Navigate around to test your app.

Step 6: Package the Android, Where's My Library App

After testing in the emulator, you will want to package your app for distribution on the Android platform to make it available to the public and your patrons. App Inventor makes this process relatively easy. From the Designer view, visit your "My Projects" list. You should see a list of the projects we have been working on. Click the checkbox next to our project, and then select the "More actions" dropdown. Within the dropdown, pick the "Download Source" to get a .zip file of all the files in your app. See Figure 5.5 if you need a visual reminder of where to find the "My Projects" list.

Step 7: Distribute the Android, Where's My Library App in the Android Market

To make your app publicly available, you will need to visit the Android Market (https://market.android.com/publish/) and sign in using a

Google Account. You will be asked to create a developer profile, pay a $25.00 registration fee, and to agree to the "Android Market Developer Distribution Agreement." Once you have finished registration, you will be able to submit your app by uploading your finished .zip file to the Android Market for review.

▶ BUILD A MOBILE WEBSITE FROM SCRATCH

View the demo here: http://www.alatechsource.org/techset/; filename = mobile-iui
Download associated files here: http://www.alatechsource.org/techset/; filename = iui-0.31.zip

Mobile website and app generators offer everyone the opportunity to create a mobile view of their library by using a rich-text or WYSIWYG editor interface. In this recipe, we will use the iui mobile site generator tool (http://www.hiddenpeanuts.com/msg/) created by Chad Haefele, the Emerging Technologies Librarian at the University of North Carolina at Chapel Hill Library.

Our finished project will look something like Figure 5.6. Let's get started.

Step 1: Download the iui Framework (Version 0.31)

We begin by downloading the iui mobile website framework from http://code.google.com/p/iui/. Make sure to grab version 0.31, as this is the version that the iui mobile site generator tool is optimized for.

▶ Figure 5.6: Mobile Website Generated Using the Generator at http://www.hiddenpeanuts.com/msg/

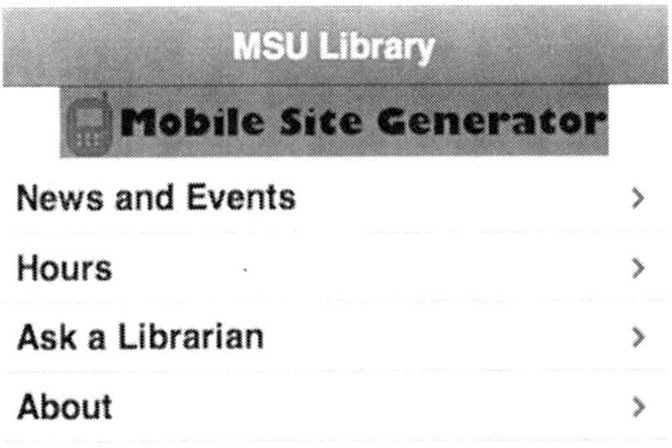

The download process is a bit tricky, so I have also made a copy of the files you will need available at http://www.lib.montana.edu/~jason/files/iui-0.31.zip. Unzip the files, and you will find a directory named "/iui/." Copy this directory and drop (or FTP) it to your public web server. The files we generate from the iui mobile site generator tool will reside in this /iui/ directory.

Step 2: Fill Out the iui Mobile Site Generator Tool Form

At this point, we can start to generate an HTML template using the iui mobile site generator tool form available at http://www.hiddenpeanuts.com/msg/. The form asks a series of questions about the title for our site, an image for the site header, and names of pages or subpages to be included in the site (see Figure 5.7).

(A stock image for the site header is available at http://www.hiddenpeanuts.com/msg/samples/minimal/msg.jpg.) You can include all types of content for the pages, but you will want to keep in mind

▶ Figure 5.7: View of Mobile Site Generator Form

Header image URL:
This is the header image for your mobile site, which will be displayed on every page. It needs to be 320x40. Here's a sample to play with: http://www.hiddenpeanuts.com/msg/samples/minimal/msg.jpg (it's so terrible that I can guarantee nobody will ever steal it)

Main page title:
This is the title of your mobile site's main page. Because of the way iUI navigation is displayed, I suggest something simple and short like 'Home'.

These items are displayed on your mobile site's main page. They're links to subpages, so should probably be things like 'hours', 'locations', 'catalog', etc. Whatever you like. The screenshot gives you an idea of what they'll look like.

Item 1:
Item 2:
Item 3:
Item 4:
Item 5:
Item 6:
Item 7:
Item 8:

Mobile Site Generator
Item 1
Item 2
Item 3
Item 4
Item 5
Item 6
Item 7
Item 8

These are the subpages that will be linked to on the *Item 1* subpage that you named above. If a user picks *Item 1*, they'll be taken to a page displaying the links you specify below. So, for example: If Item 1 above says 'hours', then these might say 'Hours for branch A', 'Hours for branch B', etc. Again, whatever you like. If you leave these boxes blank, this page can be used for text content instead.

Item 1A:
Item 1B:
Item 1C:
Item 1D:

Mobile Site Generator
Item 1A
Item 1B
Item 1C
Item 1D
Put your text content here

common mobile webpages such as Hours, Ask a Librarian, or News and Events as possible pages. And, in this case, be sure to create an Hours page using the mobile site generator form, as we will be using it as an example in Step 4.

Step 3: Get the HTML Code and Paste It into index.html in the /iui/ Directory

Once you have filled out the form hit the "Submit" button, and you will get some generated HTML code with your supplied values. Open up your text editor, and create a new HTML page where you will copy and paste this code, name it "index.html," and place it into your /iui/ directory that you created in Step 1. This will now serve as your starter template for your mobile site. The iui framework uses a single page design method where all the page views for your site are located on the index.html page. Page views for iui are marked up adding special id values to the HTML elements. For example, the following markup—

```
<ul id="AskaLibrarian">
```

—identifies the section of the index.html page where our "Ask a Librarian" page content will go. Consequently, all of our page sections will be marked up on our index.html page.

Step 4: Create Markup and Content for the Blank Template

One of our last steps is creating the unique markup and content for the starter template we generated. One of the pages created in our starter template or that you generated in Step 2 was for the hours at the library. Adding the hours information is very simple. Open up index .html in your text editor, and find the following HTML comment:

```
<!--Begin Hours subpage:-->
```

These are the HTML tag "markup hooks" that were created by the iui mobile site generator tool, and they provide our cues for where to drop our new content. A little further down in the markup, you will see a table tag that looks like this:

```
<table>
  <tr>
    <td>
    Put your text content here
    </td>
  </tr>
</table>
```

To complete the Hours page, you just need to replace the placeholder text "Put your text content here" with your hours HTML content. A final page might look like this:

```
<table>
  <tr>
    <td>
      <ul>
        <li>Monday - Thursday: 7am to 9pm</li>
        <li>Friday: 7am to 5pm</li>
        <li>Saturday: 10am to 5pm</li>
        <li>Sunday: 10am to 9pm</li>
      </ul>
    </td>
  </tr>
</table>
```

Repeat the process for all of your other page content sections, and then save the index.html file. Use the <!—Begin YOURPAGENAME subpage:—> HTML comment tags as your guide for where to start adding your page view content.

Step 5: Rename the /iui/ Directory, and Deploy the Mobile Website

The final step is renaming the /iui/ directory to a mobile-friendly name or convention. Until now, we have been using the /iui/ directory for development purposes, but with our development done we want to make it publicly addressable in a URL. Rename the /iui/ directory to /m/ (or whatever makes sense for your library mobile site). With that change, we can use the /m/ directory as a link from our library homepage to a mobile version of the site.

A Quick Note about Other Mobile Website Generators

The list of mobile site generator tools is growing every day. Google Sites now provides mobile templates (http://www.google.com/sites/help/intl/en/mobile-landing-pages/mlpb.html). Using the Google Sites mobile template wizard you can create a professional-looking mobile website in no time for free. Another useful tool to keep in mind is Winksite (http://winksite.com/). You have to sign up for an account, but once you are registered you can fill out a series of forms specifying the content you want to mobilize and have a working mobile site in a matter of minutes. If you haven't done so already, you may want to check out the first project in this chapter, which covers creating a mobile website using Winksite.

▶ BUILD AN IPHONE APP

Download Project 5.3 source files here: http://www.alatechsource.org/techset/; filename = mobile-iui.zip

Download Project 5.7 source files here: http://www.alatechsource.org/techset/; filename = touch-jquery.zip

In a library with limited developer resources, development of an iPhone application using the iOS platform can be a daunting task. Setting aside questions of digital divides and which native platform to invest in, we're going to look at ways to use existing skills in our library to build an app that can be deployed on the Apple iOS platform, which includes the iPhone and the iPod Touch. Tools, like PhoneGap and Sencha Touch, now exist that allow a developer to build a standard HTML mobile site and compile that site into a native app version for iOS devices.

For our purposes, we are going to use the files from Project 5.3, "Build a Mobile Website from Scratch," as our source code. You could also use the HTML, CSS, and Javascript files from the jQuery Mobile website app from Project 5.7, "Build a Mobile Site Using JavaScript Frameworks." In the current project, we are going to take the iui code from Project 5.3 and place the files within the PhoneGap framework to create an app for the iPhone. All of the heavy lifting and conversion to iOS programming will be handled by PhoneGap. Note that the files from Project 5.3 or the jQuery files from Project 5.7 need some customization for your local library information and search settings. If you have plans to deploy this application, please refer to those projects for the full details of customization and finish those projects before working through this project. The finished project files from Project 5.3 or 5.7 will provide you with a working HTML mobile website that you can then use with PhoneGap. Much like our Android example in Project 5.2, we are showing you the routine and processes you would use to create an iPhone app. In the end, we will have a demo app that can be packaged and then submitted for approval to the iTunes App Store.

An important note about requirements: Because we are developing for the Apple iOS platform, this particular project requires an Intel-based computer running Mac OS X Snow Leopard (10.6) with Xcode and the iOS SDK (Software Development Kit) installed. There are documented workarounds for getting Xcode and the iOS SDK to run on Windows and PC machines. See the tutorial "Installing iOS SDK and Xcode on Windows 7" for an overview (http://ipodtoucher55.blogspot.com/2010/12/installing-ios-sdk-and-xcode-on-windows.html). The Windows and PC workarounds can be complex and usually involve running a version of MAC OS X on your PC, but if this is your only option it is worth investigating.

Step 1: Register as a Developer, and Install Xcode, iOS SDK, and PhoneGap

Our first steps will be registering as an Apple developer and downloading the software packages needed to build our native app:

1. Registering as a developer is free and can be done at http://developer.apple.com/programs/register/. Once you register, you will get access to tools and resources and be able to submit your app to the App Store for review.
2. Next, we download and install the tools needed to create our app. Xcode is the software development interface for building native iOS applications. The iOS SDK is included in the Xcode download. For this project we are using Xcode 3, which is available at https://connect.apple.com/cgi-bin/WebObjects/MemberSite.woa/wa/getSoftware?bundleID=20792. (The newest version of Xcode is Xcode 4, which is available at http://developer.apple.com/devcenter/ios/index.action#downloads. The Xcode 4 download does require your registering as an Apple developer for a $99 fee, but it might be worth it if your organization is going to invest in building apps for the Apple iOS platform.)
3. Once you have downloaded Xcode and the iOS SDK, unpack/unzip the files and follow the installation instructions.
4. The final part of the Step 1 process is to download and install PhoneGap. You will find the PhoneGap download at http://www.phonegap.com/download. Unpack/unzip the file and store the PhoneGap directory on your local machine. One of the real advantages of using a framework, like PhoneGap, is the ability to compile your app for multiple platforms. We are focusing on iOS development in this project, but there are similar tutorials available that show how you might compile your app for an Android device. To learn more about developing for the Google

Android platform using PhoneGap and a Windows PC, see the tutorial at http://www.phonegap.com/start#android.

Step 2: Set Up PhoneGap to Work with an iOS Project

1. Before moving forward with the Step 2 process, make sure Xcode is turned off and closed, as it will conflict with the PhoneGap customization.
2. With the PhoneGap directory unzipped and saved to your local machine, open up the directory and notice the multiple directories with recognizable mobile platform names: Android, Blackberry, iOS, and so forth. Each of these directories contains code or installer packages to help you create a project for the specific platform.
3. We are looking to build an app for iOS, so open up the iOS directory and run the PhoneGapInstaller.pkg. Once the installer finishes the install routine, we are ready to start building our app using Xcode.

Step 3: Create a New Xcode Project

1. Create a folder on your local machine named "test-mobile."
2. Launch Xcode. Then under the File menu, select "New Project."
3. Navigate to the User Templates section in the left pane, and select "PhoneGap."
4. In the right pane, select "PhoneGap-Based Application."
5. Select the "Choose" button in the bottom right corner, name your project "test-mobile," and choose the location (in this case, use the local folder we created named "test-mobile") where you want the new project to be.

Once you have finished, Xcode will pop up a developer's window that looks like Figure 5.8.

Step 4: Place HTML Files into the PhoneGap Project

1. When we created and named our Xcode project in Step 3, we created a local directory named "test-mobile." (It can be named whatever you want as long as you remember the name and location on your local machine.) Navigate to that local directory, open it, and look for the www directory.
2. We are going to be using the existing iui mobile site application files we built in Project 5.3 to provide files for our PhoneGap

▶ **Figure 5.8: Xcode Project Window Showing PhoneGap "www" Directory**

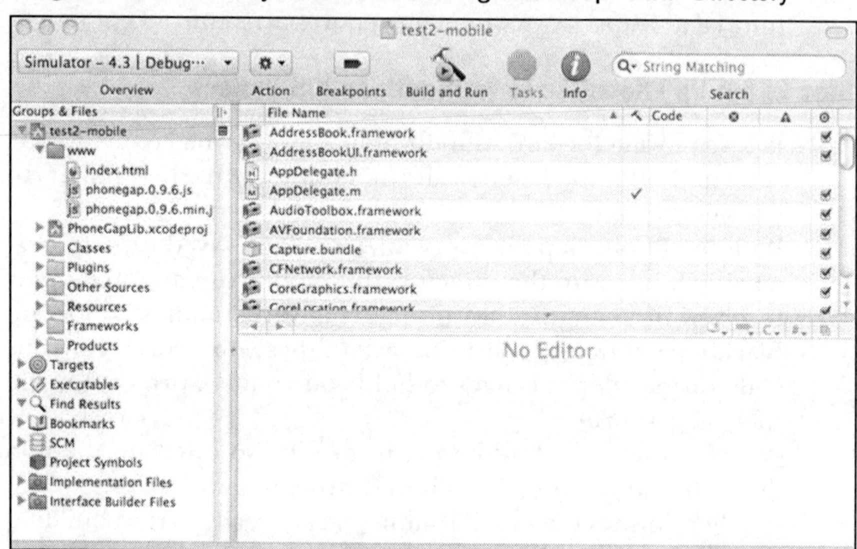

project. If you have finished Project 5.3 or 5.7, use the customized files you built while completing those projects. If not, download the source files from Project 5.3 at http://www.lib.montana.edu/~jason/files/mobile-iui.zip. Add the unzipped files from Project 5.3 to the contents of the www directory. We are adding the HTML, CSS, and Javascript files that PhoneGap will compile into our native iOS application. Don't worry about overwriting the existing index.html file in the www directory. It is merely a placeholder generated by Xcode when we created the project. Be sure to leave the phonegap.js files in place, as they will create the native application behaviors for our compiled app.

Step 5: Set Up the iOS Emulator in Xcode

Now that all the files from Project 5.3 have been placed in the www directory, we can run some tests to see how our app will perform when it is live on an iPhone. Software development programs use tools called "emulators" to mimic how a program will work when it has been released as working software. In Step 5, we are going to use Xcode's emulator to check how our app will work:

1. Once all the files from our Project 5.3 have been placed in the www directory, refresh the Xcode view by opening and closing www directory in the Xcode project view (see Figure 5.8 for reference).

2. Once refreshed, you should see index.html, a "meta" directory, and the phonegap.js scripts from your Xcode project view. We can now set up our iOS emulator to test how our app will perform on Apple's native platform.
3. In the top right corner of our Xcode project view, you will see a dropdown with the label "Simulator." (Refer to Figure 5.8 if confused.) From the dropdown, you will see a list of Apple iOS platforms; choose one of the iPhone platforms, as we want to test our app within the Apple smartphone environment.
4. Next, we are going to run the emulator, and we will see our app as it will behave on an iPhone. To launch the emulator, click on the "Build and Run" icon in the top navigation options of the Xcode project window. The iOS Simulator will take a little time to compile the app the first time it is run, but, once it is finished, you will see your app in a native iPhone frame (see Figure 5.9).

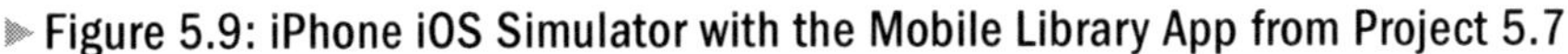

▶ Figure 5.9: iPhone iOS Simulator with the Mobile Library App from Project 5.7

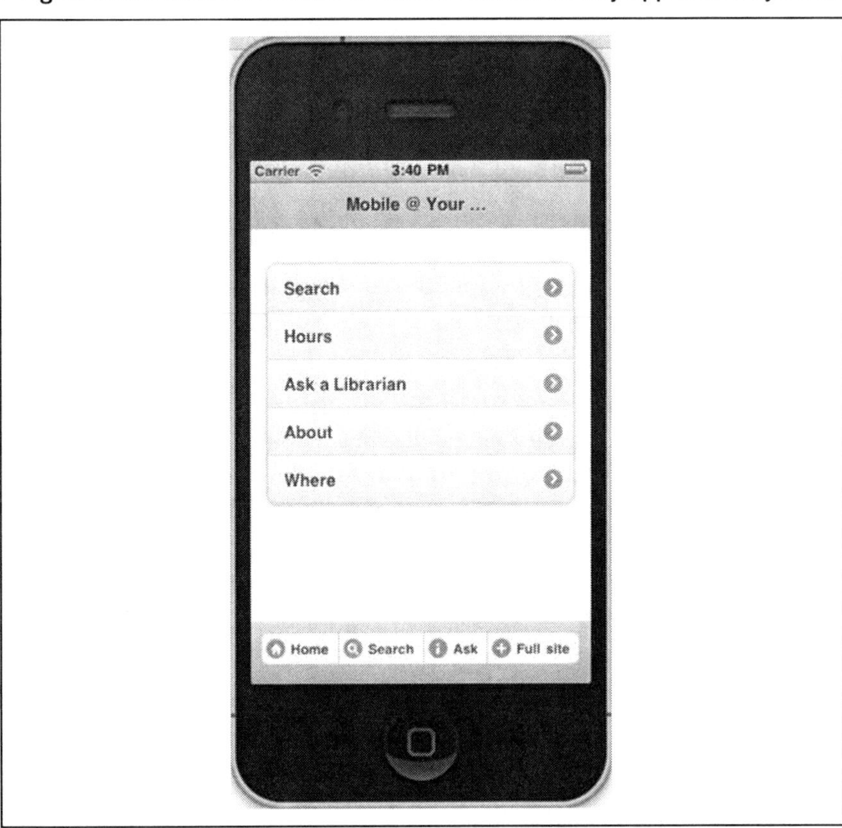

This is a live emulator, and you should click around your app to see how it performs. Pay particular attention to the primary actions of your application in this environment.

Step 6: Finalize PhoneGap Files for iOS

If you are satisfied with your tests in the emulator, our next steps involve preparing the app icons for the iPhone homepage and the initial screen that will appear as the application is opening and loading on the iPhone:

1. Within the Xcode project view, find the Resources directory. (See Figure 5.8 for reference.) Within the Resources directory, you will find an "icons" directory that holds the thumbnail images that will be the default images for your app on the iPhone homepage.
2. Another directory named "splash" holds all the default splash screen images that will be displayed while your app is loading.
3. By going to your local PhoneGap project directory (we named it "test-mobile"), you can open these files in your preferred image editor to get the dimensions and replace all of the images with your default thumbnails and splashscreen images. Note: Replacing images is not required, but in order to distinguish or market your app with an identity, such as your library logo, it is a best practice.

Step 7: Package Files, and Submit/Register Your Application

Our final step is to package our files and register the application for review by Apple. We will use the Xcode Project Viewer to complete this process:

1. First, we need to get an application identifier from Apple by visiting http://developer.apple.com/iphone/manage/overview/index.action and logging in with our developer account we set up at the beginning of this project. The application identifier is similar to a serial number that Apple uses to associate you as a developer with your unique app. Once you are in your Apple account view, click on the "App IDs" link in the left navigation and look for the "New App ID" button on the right. Press the button and you should see the Create App ID screen like the one in Figure 5.10.

   Fill in all the form fields and press "Submit." When finished, you will be given an application ID. Copy it to your clipboard, and save it for a later step.

▶ **Figure 5.10: View of Apple Account When Registering Application ID**

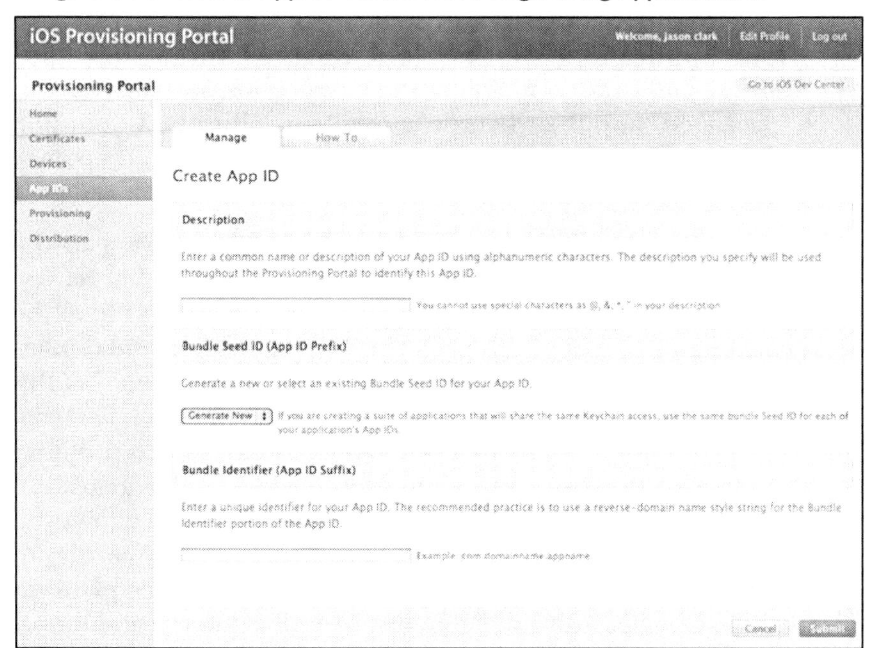

2. Once you have the identifier, open up your Xcode software, and find the project named "test-mobile" to load the project you want to work on packaging for review.
3. Next, go to the Resources directory in the Xcode Project Viewer, open up [AppName]-Info.plist (for this example it would be called "test_mobile-info-plist"), and change the Bundle Identifier to the identifier provided by Apple. Changing the identifier allows Apple to register the application as your own and gives your app access to Apple platform features like cloud file storage.
4. Make sure to change the Active SDK in the top left dropdown menu from "Simulator-5.3" to the device and SDK version we are using to build the application. In this case, we are specifying that we are deploying our application to a device and which SDK version we want our application to work on. A good choice here is "5.3 and Device," so be sure to set the top left dropdown with these values.
5. Under the Build menu in Xcode, select "Build and Archive" in your Xcode project window header. After the build is completed, you should now have an archived or "packaged" app within Xcode that can be submitted to iTunes Connect.

6. Log in to iTunes Connect (http://itunesconnect.apple.com/) using your developer ID and credentials. Click on the "Manage Your Applications" option, look for the "Add a New App" button, and click on it. Follow the instructions, and upload your "packaged" app file when prompted. Once finished, your app will go under review, and you will be notified of its acceptance into the Apple App Store.

As you can see, the whole process of building and deploying an app can be very complex. After completing the steps outlined here, you now have a sense of the routines and processes involved. One of the most challenging pieces of this project is the registration and deployment of the application. If you are having trouble with Step 7 of this project, there is a "beta" (invite only as of this writing) service from the creators of PhoneGap named "PhoneGap Build," which promises to take some of the pain out of the deployment and registration process (http://build.phonegap.com/). PhoneGap Build allows you to write your app in HTML, CSS, and JavaScript, upload your files to the PhoneGap Build servers, and then choose which mobile platform (Apple iOS, Google Android, Palm, Symbian, BlackBerry, and more) you want your application to work on. When finished, you receive your app store–ready application that you can submit to the required mobile platform. The service is set to remain free for open source projects, but a minimal price tag could be completely worth it. At any rate, you now have the full picture of iPhone app development using PhoneGap, and you should feel confident in experimenting with the possibilities.

▶ MOBILIZE YOUR LIBRARY'S CATALOG

View the demo here: http://www.alatechsource.org/techset/; filename = worldcat-local-search

Download the files here: http://www.alatechsource.org/techset/; filename = worldcat-local-search.zip

One of the primary interests of libraries making the move to mobile is to provide a catalog search view for the mobile setting. In this project, we will look at how to use the WorldCat Search API and the HTML5 JavaScript geolocation API to build a search that provides views of your library catalog data based on WorldCat holdings. (API stands for application programming interface. In simple terms, an API is a set of rules or functions that allow one computer system to access data and services from another remote computer system.) This mobile web app will search against the WorldCat holdings index, identify local libraries around you that have the item you are looking for, and link into your catalog record if your library has the item. In this particular project, we will not be writing code from scratch. The steps detailed here will focus on downloading the necessary files for the application and editing certain pieces of code to create a custom mobile catalog search for your library. In the interest of teaching about how the app was built and works, many of the steps are instructive or explanatory and will not require specific edits. When necessary, edits and changes are clearly indicated. Requirements for this app include PHP, JavaScript, HTML, and CSS, so you will want to make sure your local systems support these technologies. Figure 5.11 shows a screenshot of what a search results page will look like in the app we'll be customizing.

▶ Figure 5.11: Search Results Screen for WorldCat Search API Mobile Catalog

Mobile @ Your Library
Search
Explore
About
Search Your Library
keyword, name, title...
977 matches for your query "zeitoun"
Next
Zeitoun
http://worldcat.org/oclc/332257182
Author(s): Eggers, Dave.
Summary: In the wake of Hurricane Katrina, longtime New Orleans residents Abdulrahman and Kathy Zeitoun are cast into an unthinkable struggle with forces beyond wind and water. In the days after the storm, Abdulrahman traveled the flooded streets in a secondhand canoe, passing on supplies and helping those he could. A week later, on September 6, 2005, Zeitoun abruptly disappeared-- arrested and accused of being an agent of al Qaeda.
OCLC ID: 332257182
ID: urn:ISBN:9781934781630
Citation (Worldcat full site)
Find local libraries
Search local catalog
Next
Full site
Subscribe

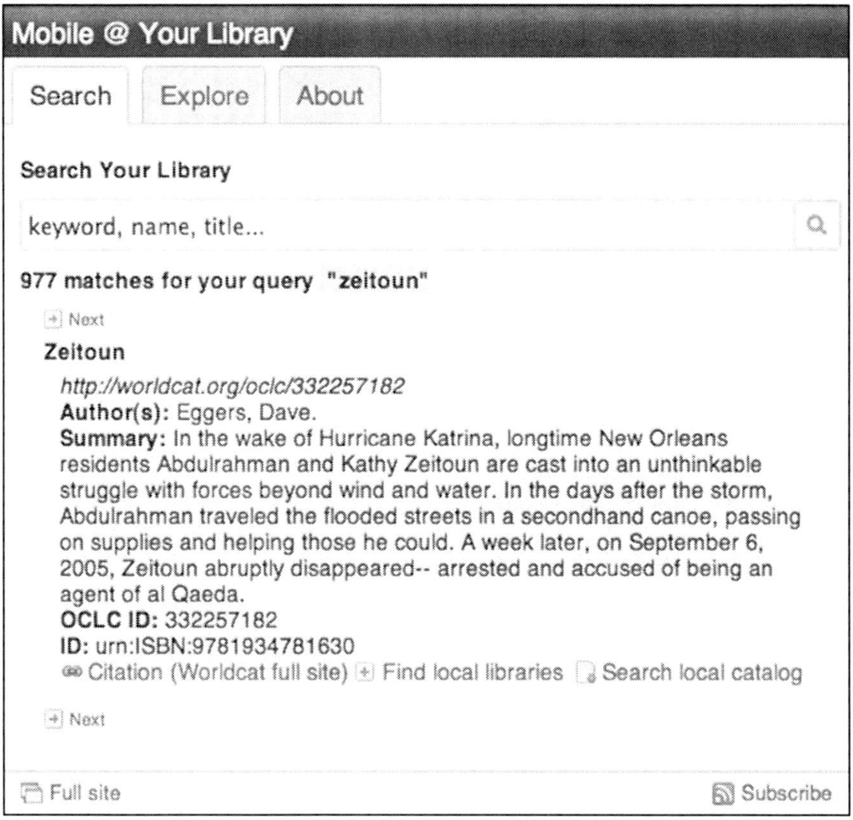

Step 1: Get an Account for the WorldCat Search API

The data source for our app is the WorldCat holdings index. We are going to run queries against this index to create specific views of data for our application. To make these queries for WorldCat data, you will need to register for an account to use the WorldCat Search API. Your account will provide you with an API key that will allow you to access the WorldCat Search API web service. To sign in using an existing account or to create a new one, visit https://worldcat.org/config/SignIn.do. Once you have logged in, look for the "Request an API Key" link in the left navigation. Record your API key in a safe place, as you will need it later. (Full details, documentation, and the terms of service for the WorldCat Search API can be found at http://www.oclc.org/developer/services/WCAPI.)

Step 2: Download the Mobile Catalog Search Template Files

As mentioned in the introduction, we will be downloading a "pre-made" working mobile catalog app and customizing it for your local catalog. Get the files at http://www.lib.montana.edu/~jason/files/worldcat-local-search.zip. Unzip the files and open up the following files in your text editor:

- index.php
- /views/local.php
- /views/search.php
- /views/where.php
- /meta/styles/m.css

Step 3: Build the Display Interface, and Name the Application

Once we have an account and API key and the files open, we can get started building the app in earnest. It is useful to think of our app as having two functional components: a display interface for interactions and a processing function where data is requested and parsed. The public display interface will be built with HTML, CSS, and JavaScript, while the "back end" processing functions will use PHP to make different views of data appear during different app functions. To illustrate these separate layers of the application, let's take a closer look at the initial page view for our app. At this point, we are just explaining what lines in the files are doing in order to help you learn how the search and display pages are working. No need for edits, just yet. Our interface layer is a search form built with the following HTML markup in the source file of /views/ search.php:

```
<form id="searchBox" method="get" action="./index
 .php?view=search">
<input type="hidden" name="lat" id="lat" value="" />
<input type="hidden" name="lng" id="lng" value="" />
<fieldset>
<label for="q">Search</label>
<input type="text" maxlength="200" name="q" id="q"
 tabindex="1" value="keyword, name, title..."
 onclick="if (this.value == 'keyword, name, title...')
 { this.value = ''; }" onblur="if (this.value == '')
 { this.value = 'keyword, name, title...'; }" />
<button type="submit" class="button">Search</button>
</fieldset>
</form>
```

This code makes a search form appear in the browser, and it is styled for display using the following CSS styles from the /meta/styles/m.css file:

```
input,select {vertical-align:middle;border-width:0;}
fieldset {border:0 none;}
form#searchBox {position:relative;width:100%;height:
 30px;background-color:#fff;border:1px solid#ddd;
 border-radius:3px;-moz-border-radius:3px;-webkit-
 border-radius:3px;}
#searchBox label {display:none;}
#searchBox input {position:absolute;top:3px;left:0;
 margin:0 3px;padding:0;width:85%;line-height:25px;
 height:25px;font-size:1.4em;border:0;-webkit-
 appearance:none;}
#searchBox button {position:absolute;top:0;right:0;
 cursor:pointer;margin:0;padding:0;width:30px;height:
 30px;float:left;background: url('../img/search.gif')
 no-repeat 50% 50%;text-indent:-999em;border:1px solid
 #ddd;border-radius:3px;-moz-border-radius:3px;-webkit-
 border-radius:3px;}
```

And with that, we have our HTML markup and CSS styles for our display interface. A user can start typing queries into our search form and get results presented to them in a list. But, we need one more change to customize the application—a unique title or display name. Go to index.php in your text editor, and look for lines 2–4. The lines should look like this:

```
//set value for title of page
$pageTitle = 'Mobile';
$subTitle = '@ Your Library';
```

These lines are the place where you set the display name for your app that will appear at the top of the screen. Add your specific title in the $pageTitle and $subTitle variables, replacing 'Mobile' and '@ Your Library' with your unique title. You will want to be mindful of the limited space a mobile device will display. A suggested title might look like this:

```
//set value for title of page
$pageTitle = 'MSU Library Search';
$subTitle = 'Mobile;
```

Two other customizations should also be made here. We want our links in the footer of our page to point at our full desktop website and a local feed URL to allow mobile users to access our full site and full feeds if they choose. Look for the following HTML markup near the bottom of the page in index.php:

```
<a accesskey="4" class="site" title="full site" href=
 "/">Full site</a>
<a accesskey="5" class="feed" title="feed for
 collection" href="../feed.xml ">Subscribe</a>
```

In the first href= value, replace the "/" with the link to your full desktop website. In the second href= value, replace the "../feed.xml" with a link to a local library feed URL (e.g., your library news and events RSS feed). After you have settled on your title, your link to your full site, and your feed URL, save index.php into the same directory where you unzipped the file in Step 2.

Step 4: Build the Processing Layer, and Set Your WorldCat Search API Key and Your OCLC Library Code

The processing layer of our application is built around the goal of getting data or results from the WorldCat API. Take a closer look at the action on our form from Step 2:

```
<form method="get" action="./index.php?view=search">
```

When someone enters a search term and touches the search button, the action index.php?view=search is fired, and here is where our processing layer for the app starts to work. In this case, our processing layer is a PHP script that is part of the file located at /views/search .php. In Step 1, we signed up for the WorldCat API account and got our developer's key, and we are now going to use those credentials to enable our script to make requests to the WorldCat Search API. Lines 2–13 in the /views/search.php list our default values and settings for our search processing functions:

```
//set default value for Worldcat API key
$key = isset($_GET['key']) ? trim(strip_tags(urlencode
  ($_GET['key']))) : 'YOUR-WORLDCAT-KEY';
//set default value for query
$q = isset($_GET['q']) ? trim(strip_tags(urlencode
  ($_GET['q']))) : null;
//set default value for latitude
$lat = isset($_GET['lat']) ? $_GET['lat'] : null;
//set default value for longitude
$lng = isset($_GET['lng']) ? $_GET['lng'] : null;
//set default value for library collection to search
$library = isset($_GET['library']) ? trim(strip_tags
  ($_GET['library'])) : 'YOUR-OCLC-LIBRARY-CODE';
```

To make the script work for your local library, there are two pieces of information that you must add to these credentials. Open up the file named "search.php" in the Views folder, and find lines 2–13 at the top of the script. First, you must add your WorldCat API key value that you received when you signed up for your developer's account to the $key variable. Adding your key will validate that you are a registered user of the WorldCat APIs and you are allowed to query the OCLC web service for data. Second, you must add your OCLC library code to the $library variable. The $library variable is one of the most essential edits you will make to the existing source code for this application. Adding your OCLC library code here will allow the script to add links to your local catalog and make the app work as a local catalog search. You can get your OCLC code by visiting http://www.oclc.org/contacts/libraries/. (For more information on creating local catalog links using the Worldcat API, see http://oclc.org/developer/documentation/worldcat-search-api/library-catalog-url.) When finished with these edits, save search.php to the Views folder.

Step 5: Make Requests to the WorldCat Search API

Our next step in creating a processing layer for our app is to actually use the data that someone submitted from the search form to make a request to the WorldCat Search API. The WorldCat Search API is a web service gateway, which means we can make a request using a specifically formatted URL. We send the formatted URL to the WorldCat Search

API, and we get formatted data in return. For most of this step, we will be looking at the code that makes things go and explaining what is going on. If a code change is required, it will be indicated.

In our app, we make one of our first calls to the WorldCat Search API to retrieve a list of WorldCat holdings based on a query from our search form. Our script uses the values that came from the search form to populate the $q, $lat, and $lng variables.

```
//set default value for query
$q = isset($_GET['q']) ? trim(strip_tags(urlencode
  ($_GET['q']))) : null;
//set default value for latitude
$lat = isset($_GET['lat']) ? $_GET['lat'] : null;
//set default value for longitude
$lng = isset($_GET[‘lng’]) ? $_GET['lng'] : null;
```

$_GET tells our script to use the values in the URL to assign data to the variables. These variable values were pushed into the URL when our search form was submitted. Another closer look at the search form should help illustrate this:

```
<form id="searchBox" method="get" action="./index
 .php?view=search">
...
<input type="hidden" name="lat" id="lat" value="" />
<input type="hidden" name="lng" id="lng" value="" />
<input type="text" name="q" id="q" value="keyword... "/>
...
<button type="submit" class="button">Search</button>
</form>
```

When the form is submitted, all of the <input /> values (lat, lng, q) become part of the URL, and our script at /views/search.php uses them to create the $q, $lat, and $lng variables. These variables are combined and formatted as a second URL that will be pushed forward to the WorldCat Search API in the following lines (lines 32–43) of the /views/search.php file:

```
//set base url for our opensearch request to Worldcat
  Search API
$base = 'http://www.worldcat.org/webservices/catalog/
  search/worldcat/opensearch?';

$params = array(
  'q' => $q,
  'format' => 'atom', //type of format to output
```

```
    'cformat' => 'mla', //append citation format
    'start' => $start, //starting number for results to
      return
    'count' => $limit, // optional argument supplies
      number of results to return
    'wskey' => $key, //Worldcat API key
    //all possible options are documented at http://
      worldcat.org/devnet/wiki/SearchAPIDetails
);
```

When pieced together, the raw URL request looks like this:

```
http://www.worldcat.org/webservices/catalog/search/
  worldcat/opensearch?q=obama
&format=atom
&cformat=mla&start=1
&count=10&wskey=YOUR-API-KEY-HERE
```

Let's unpack the parameters in this URL, as this will help us understand how one communicates specific requests to the WorldCat Search API using a URL. The base URL for the web service is http://www.worldcat.org/webservices/catalog/search/worldcat/open search?. q= is our query, format is the format of data to return, cformat is the citation format to return, start is the first record result number to return, and count is the number of record results to return. Once the request is in order, our script will pass the URL to the WorldCat Search API. Again, in the interest of instruction, we will look at the response from the WorldCat Search API to understand how the whole process works together. Here's the formatted data that the request delivers:

```
<?xml version="1.0" encoding="UTF-8" standalone="no"?>
<feed xmlns="http://www.w3.org/2005/Atom" xmlns:URLEn-
  coder="java://java.net.URLEncoder" xmlns:xsi="http:
  //www.w3.org/2001/XMLSchema-instance" xmlns:diag=
  "http://www.loc.gov/zing/srw/diagnostic/" xmlns:
  oclcterms="http://purl.org/oclc/terms/" xmlns:dc=
  "http://purl.org/dc/elements/1.1/" xmlns:opensearch=
  "http://a9.com/-/spec/opensearch/1.1/">
<title>OCLC Worldcat Search: obama</title>
<id>http://worldcat.org/webservices/catalog/search/
  worldcat/opensearch?q=obama&amp;start=1&amp;count=1&
  amp;format=atom&amp;wskey=B3F6fY0fdaYyWFaU2a5a25QD28
  BsxH6H8wZnViTESKxZZBR7Fg71nC0V6IeXa78EKAYsGzhMAyYy
  Eihv</id>
<updated>2011-06-06T18:50:59-04:00</updated>
```

```
<subtitle>Search results for obama at http://worldcat
 .org/webservices/catalog</subtitle>
<opensearch:totalResults>7094</opensearch:totalResults>
<opensearch:startIndex>1</opensearch:startIndex>
<opensearch:itemsPerPage>1</opensearch:itemsPerPage>
<opensearch:Query role="request" searchTerms="obama"
 startPage="1"/>
<link rel="alternate" href="http://worldcat.org/web
 services/catalog/search/worldcat/opensearch?q=obama&
 amp;start=1&amp;count=1&amp;wskey=B3F6fY0fdaYyWFaU2a5
 a25QD28BsxH6H8wZnViTESKxZZBR7Fg71nC0V6IeXa78EKAYsGzh
 MAyYyEihv" type="text/html"/>
<link rel="self" href="http://worldcat.org/webservices/
 catalog/search/worldcat/opensearch?q=obama&amp;start=
 1&amp;count=1&amp;wskey=B3F6fY0fdaYyWFaU2a5a25QD28BsxH
 6H8wZnViTESKxZZBR7Fg71nC0V6IeXa78EKAYsGzhMAyYyEihv&
 amp;format=atom" type="application/atom+xml"/>
<link rel="first" href="http://worldcat.org/webservices/
 catalog/search/worldcat/opensearch?q=obama&amp;start=1
 &amp;count=1&amp;wskey=B3F6fY0fdaYyWFaU2a5a25QD28BsxH6
 H8wZnViTESKxZZBR7Fg71nC0V6IeXa78EKAYsGzhMAyYyEihv&amp;
 format=atom" type="application/atom+xml"/>
<link rel="next" href="http://worldcat.org/webservices/
 catalog/search/worldcat/opensearch?q=obama&amp;start=
 2&amp;count=1&amp;wskey=B3F6fY0fdaYyWFaU2a5a25QD28BsxH
 6H8wZnViTESKxZZBR7Fg71nC0V6IeXa78EKAYsGzhMAyYyEihv&
 amp;format=atom" type="application/atom+xml"/>
<link rel="last" href="http://worldcat.org/webservices/
 catalog/search/worldcat/opensearch?q=obama&amp;start=
 7094&amp;count=1&amp;wskey=B3F6fY0fdaYyWFaU2a5a25QD28
 BsxH6H8wZnViTESKxZZBR7Fg71nC0V6IeXa78EKAYsGzhMAyYyEihv
 &amp;format=atom" type="application/atom+xml"/>
<link rel="search" type="application/opensearch
 description+xml" href="http://worldcat.org/webservices/
 catalog/opensearch.description.xml"/>
<entry>
<author>
<name>Obama, Barack.</name>
</author>
<title>The audacity of hope : thoughts on reclaiming
 the American dream</title>
<link href="http://worldcat.org/oclc/71312726"/>
<id>http://worldcat.org/oclc/71312726</id>
<updated>2011-02-11T11:51:28Z</updated>
```

```
<content type="html">&lt;p class="citation_style_MLA"&gt;
 Obama, Barack. &lt;i&gt;The Audacity of Hope: Thoughts
 on Reclaiming the American Dream&lt;/i&gt;. New York:
 Crown Publishers, 2006. Print. &lt;/p&gt;</content>
<summary>Senator Obama calls for a different brand
 of politics—a politics for those weary of bitter
 partisanship and alienated by the "endless clash of
 armies" we see in Congress and on the campaign trail; a
 politics rooted in the faith, inclusiveness, and nobility
 of spirit at the heart of our democracy. He explores
 those forces--from the fear of losing, to the perpetual
 need to raise money, to the power of the media--that can
 stifle even the best-intentioned politician. He examines
 the growing economic insecurity of American families, the
 racial and religious tensions within the body politic, and
 the transnational threats--from terrorism to pandemic--
 that gather beyond our shores. And he grapples with the
 role that faith plays in a democracy. Only by returning
 to the principles that gave birth to our Constitution,
 he says, can Americans repair a broken political
 process, and restore to working order a government
 dangerously out of touch with millions of ordinary
 Americans.--From publisher description.</summary>
<dc:identifier>urn:ISBN:0307237699</dc:identifier>
<dc:identifier>urn:ISBN:9780307237699</dc:identifier>
<dc:identifier>urn:LCCN:2006028967</dc:identifier>
<oclcterms:recordIdentifier>71312726</oclcterms:record
 Identifier>
</entry>
</feed>
```

The structured data returned from the WorldCat Search API is XML. You can see that all the information we might want to display about an item is available: links, summaries, titles, authors, identifiers, and so forth. Our processing layer now needs to "cherry pick" the pieces of data we want to display in our search results. Now that we have seen how our script makes a request, we need to tell our script to do something with the data that was returned.

Step 6: Parse the WorldCat Opensearch XML File to Create a Results View

Our next step in building the processing layer is to create our search results view by picking our way and choosing the pieces of data from the WorldCat Opensearch XML file that we would like to display. Once again, this step will focus on explaining the code to make sure you have a full sense of how the application works. The PHP code that creates our search view by selecting pieces of data from the XML file shown in Step 5 begins around line 73 in the /views/search.php file:

```
//build request, send to Worldcat Search API
$request = simplexml_load_file($base.http_build_query
  ($params));

//create xml object(s) out of response from Worldcat
  Search API
$data = $request;
...
//check for results, parse xml elements, and display as html
  echo '<ul class="result">'."\n";
    foreach ($data->entry as $result) {
      //prepare opensearch namespace for parsing
      $oclc = $result->children('http://purl.org/
          oclc/terms/');
      //prepare opensearch namespace for parsing
      $dc = $result->children('http://purl.org/dc/
          elements/1.1/');
      //display the info to user
      echo '<li>'."\n";
      echo '<h3>'.$result->title.'</h3>'."\n";
      echo '<p>'."\n";
      echo '<em>'.$result->id.'</em><br />'."\n";
      if (strlen($result->author->name) > 3) { echo
          '<strong>Author(s):</strong> '.$result->
          author->name.'<br />'."\n"; }
        elseif (strlen($result->author->name) < 3) { echo
            '<strong>Author(s):</strong> Unknown
            <br />'."\n"; }
      if (strlen($result->summary) > 3) { echo
          '<strong>Summary:</strong> '.$result->
          summary.'<br />'."\n"; }
        elseif (strlen($result->summary) < 3) { echo
            '<strong>Summary:</strong> Not available
            <br />'."\n"; }
      echo '<strong>OCLC ID:</strong> '.$oclc->record
          Identifier.'<br />'."\n";
      echo '<strong>ID:</strong> '.$dc->identifier[0].
          '<br />'."\n";
```

```
        //echo '<strong>Cite This: </strong>'.$result->
            content.'<br />'."\n";
        echo '<a class="citation" href="'.$result->id.'">
            Citation (Worldcat full site)</a>'."\n";
        echo '<a class="expand" href="./index.php?view=
            where&id='.$oclc->recordIdentifier.'&lat='
            .$lat.'&lng='.$lng.'&title='.urlencode
            ($result->title).'">Find local libraries
            </a>'."\n";
        echo '<a class="download" href="./index.php?view=
            local&id='.$oclc->recordIdentifier.
            '&library='.$library.'&title='.urlencode
            ($result->title).'">Search local catalog
            </a>'."\n";
        echo '</p>'."\n";
        echo '</li>'."\n";
    }
echo '</ul>'."\n";
```

Some quick notes about what is happening in this code. The $request variable is where we are formatting the URL request and sending it to the WorldCat Search API. When the response from the API is returned, we wrap the request in the $data variable and then start to traverse through the XML document. We know there are certain pieces of data that we need for our search result display, and our last step here is to start picking the parts of the XML file that we will need. Our search result display will use the <ul> (unordered list) HTML markup, and that's why we use the "echo" statement to print out the beginning of our list:

```
echo '<ul class="result">'."\n";
```

Shortly after, we create a programming loop that will work through all the nodes of the XML document and retrieve the values that we need for our display. The loop starts here:

```
foreach ($data->entry as $result) {
```

And within the loop we grab the values we need, like the title, author, id, and so forth, using programming expressions:

`$result->title`	will get the item title
`$result->author->name`	will get the item author

And with that, we have created our processing layer that will show our search results display.

Step 7: Apply Geolocation Using JavaScript to Set a User's Current Position

One of the primary emphases for our app is finding ways to make local views of catalog data available. By using the JavaScript geolocation API and passing some local parameter values (latitude and longitude) to the WorldCat Search API, we can help our application find the current location of the person performing the searches. At first, this step may seem peripheral, but it allows for an important fallback if a user searches for an item that is not part of your local catalog. By tracking the user's location, we will be able to guide them to local libraries that have the item. When you visit our app for the first time, you will be prompted with a dialogue box like the one shown in Figure 5.12.

The prompt shown in Figure 5.12 is a result of our app making a call to the W3C Geolocation API (http://dev.w3.org/geo/api/spec-source.html). This API allows us to find specific positions of devices

▶ **Figure 5.12: View of Local Catalog Holdings for Item Search with Geolocation Prompt**

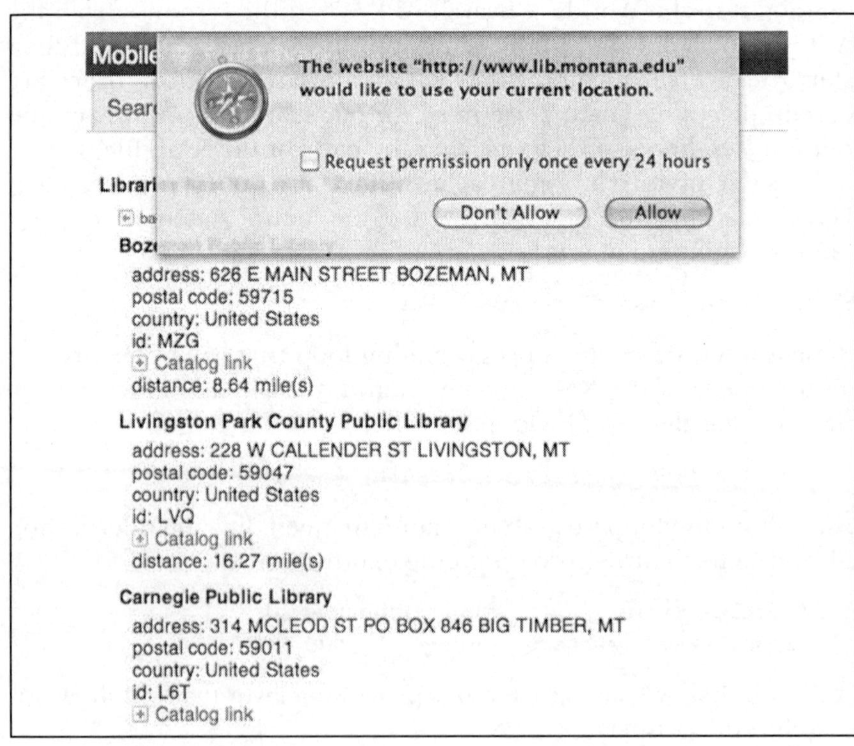

according to their locations on telecommunication networks. A typical geolocation script (and our script in this case) will look like this:

```
<script type="text/javascript">
if (navigator.geolocation) {
  navigator.geolocation.getCurrentPosition(getPosition);
  // also monitor position as it changes
  navigator.geolocation.watchPosition(getPosition);
} else {
  alert('Error: Your browser doesn\'t support
      geolocation.');
}

function getPosition(position) {
  //set latitude and longitude values
  var lat = position.coords.latitude;
  var lng = position.coords.longitude;
  //print latitude and longitude values into search
      form input hidden fields
  document.getElementById("lat").value = lat;
  document.getElementById("lng").value = lng;
}
</script>
```

The JavaScript appears in the index.php file for our app (lines 33–52). The script uses the JavaScript geolocation API to find the current location of the mobile device and then places the latitude and longitude values into the search form fields before it is submitted. Our app uses these hidden input fields to pass location data along to different stages of our processing layer.

Step 8: Make Local Catalogs Display Matches for Our Searches

Our next step in building the processing layer for our application is to create the ability to run local catalog searches within a certain radius of a user's position. You will notice in our search results that we can "Find Local Libraries" that hold items from our search results. (See Figures 5.11 and 5.12 as a reference.) The PHP script that performs this task is the file located at /views/where.php. Find and open up where.php from the Views folder in your text editor, as you will need to make an important edit. The primary action of the script starts at line 32 where we set up a request to the WorldCat Search API "library locations" (see http://oclc.org/developer/documentation/worldcat-search-api/library-locations for full documentation):

```
//set base url for our library location request to
    Worldcat Search API
$base = 'http://www.worldcat.org/webservices/catalog/
    content/libraries/'.$id.'?';

$params = array(
  'lat' => $lat, //latitude bounding area
  'lon' => $lng, //longitude bounding area
  'startLibrary' => $start, //optional argument supplies
      where to place cursor for results to return
  'maximumLibraries' => $limit, //optional argument
      supplies number of results to return
  'libtype' => $type, //type of library to search where 1 =
      academic, 2 = public, 3 = government, and 4 = other
  'format' => 'json', //append output format
  //'callback' => 'libLocations',
  'wskey' => 'YOUR-WORLDCAT-KEY', //Worldcat API key
);
```

In this instance, the $id, $lat, and $lng variables are passed from our search results script (/views/search.php) to the /views/where.php script. These variables will detail what item to look for and give a proximity location to run the search within. To make this script run, you must add your unique WorldCat search API key at line 43. After you have added your API key, save the file as "where.php" to the Views folder.

Step 9: Finalize the Local Search of Your Library Catalog

If you remember in Step 4, we added our OCLC Library Code to our search result parsing and display script (/views/search.php). To finalize and activate our local search of your library catalog, we have to make one more change. You can do this by adding your API key to the script that runs a catalog search specifically for your institution located at /views/local.php. Open up local.php in the Views folder, and add your developer key to line 17 to enable the search of your local catalog. After you have made these edits, save local.php to the Views folder, and you will have finished customizing the app for your local library catalog search.

Step 10: Customize the Look and Feel of the App Using CSS, and Load the App into a Public Web Directory

A final step in customizing the app for your library might be changing colors, link styles, or any other stylistic elements to match your library

CSS styles. The files that control these rules in our template app are located at /meta/styles/m.css. Open m.css from the /meta/styles/ directory in your text editor, and make any changes you might like. Save the file when you are finished. (Note: Tread lightly on the styles that dictate layout, such as: #doc, #hd, #main, and #ft. The template has been tested across the Android, iPhone, and major smartphone platforms and works well in those settings. Significant changes to the CSS rules that govern layout could affect the display across these platforms.)

Once you are satisfied with the changes, make sure all your open files are saved, and then move the whole folder from the template app (with all your customizations and edits) to a public web directory. You now have a working local catalog search optimized for mobile platforms.

▶ MOBILIZE YOUR SITE WITH CSS

View the demo here: http://www.alatechsource.org/techset/; filename = responsive-design

Download the files here: http://www.alatechsource.org/techset/; filename = responsive-design.zip

One of the most efficient ways to prepare your site for the mobile environment is to optimize your CSS for display on mobile devices. In this project you will learn how to analyze your current HTML markup and revise your CSS styles to make your site work in the handheld environment. We're going to take a simple two-column desktop site (shown in Figure 5.13) and make it dynamically adjust to mobile and handheld orientations and sizes (shown in Figure 5.14).

The primary idea behind this transformation is to "linearize" the layout. Mobile and handheld settings rely on a single frame for viewing, and elements stacked on top of one another fit best in this limited frame. We will be applying an emerging design technique called "responsive web design" to allow our website to scale and work across multiple computing environments, specifically desktops and smartphones. In this project, we will model the routine you can go through to add mobile CSS styles to your traditional desktop website using some demo HTML and CSS files. We are focusing on the process of adding additional HTML markup and CSS styles with the idea that you will be able to apply the routine to your own library website's HTML and CSS files. Let's get started with our demo files and project.

▶ **Figure 5.13: Traditional Desktop Browser View of Site**

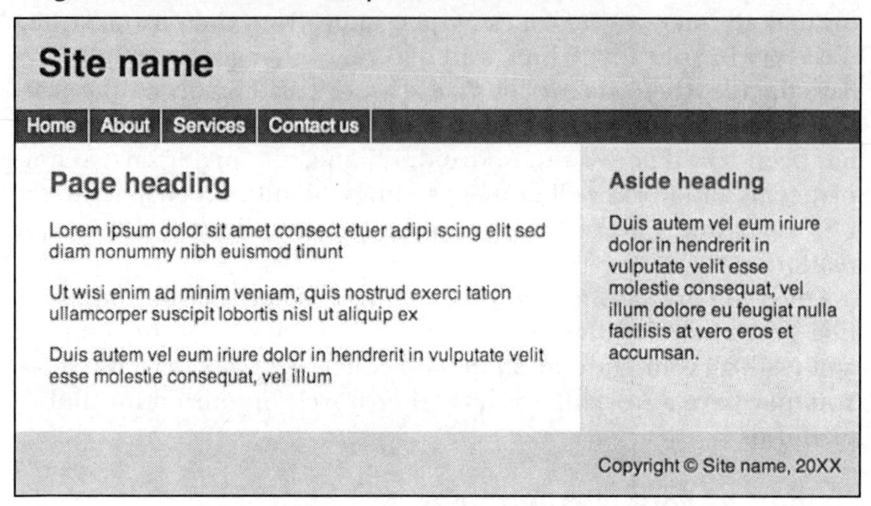

▶ **Figure 5.14: Mobile Browser View of Site**

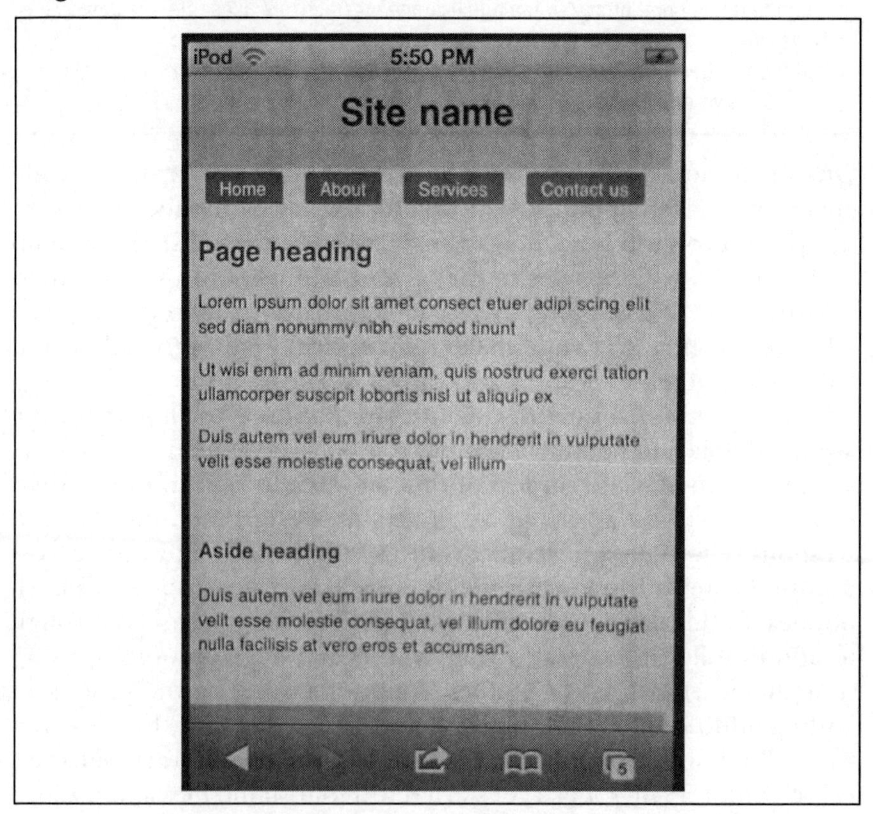

Step 1: Download the Desktop HTML and CSS Files, and Add Mobile HTML Markup

Our first step is to get the desktop HTML and CSS files from our demo at http://www.lib.montana.edu/~jason/files/responsive-design.zip. Unzip the files, and open index.html in your preferred text editor. We will be modifying minor pieces of the index.html file and extensive pieces of the CSS file as well as creating several new CSS files to bring in our "responsive web design" that will adapt to device sizes. There is one significant piece of HTML markup that is essential for our mobile optimization: adding the viewport tag. The <meta viewport tag will make our mobile design and text scale to the display of the device on which it is viewed. With the exception of the viewport tag and adding the additional links to the CSS files we will create, there is not much else to talk about here. We have some standard HTML markup that will create a two-column layout with a header and a footer. Here's the original desktop HTML markup we use along with our edit highlighted:

HTML (with abridged header and body tags)

```
...
<meta name="viewport" content="width=device-width;
 initial-scale=1; maximum-scale=1;"/>
......
<div id="container">
  <div id="header">
    <h1>
      Site name
    </h1>
  </div>
  <div id="navigation">
    <ul>
      <li><a href="#">Home</a></li>
      <li><a href="#">About</a></li>
      <li><a href="#">Services</a></li>
      <li><a href="#">Contact us</a></li>
    </ul>
  </div>
  <div id="content-container">
    <div id="content">
      <h2>
        Page heading
      </h2>
      <p>
```

```
      Lorem ipsum dolor sit amet consect etuer adipi
          scing elit sed diam nonummy nibh euismod tinunt
    </p>
    <p>
      Ut wisi enim ad minim veniam, quis nostrud exerci
          tation ullamcorper suscipit lobortis nisl ut
          aliquip ex
    </p>
    <p>
      Duis autem vel eum iriure dolor in hendrerit in
          vulputate velit esse molestie consequat,
          vel illum
    </p>
  </div>
  <div id="aside">
    <h3>
      Aside heading
    </h3>
    <p>
      Duis autem vel eum iriure dolor in hendrerit in
          vulputate velit esse molestie consequat,
          vel illum dolore eu feugiat nulla facilisis
          at vero eros et accumsan.
    </p>
  </div>
  <div id="footer">
    Copyright © Site name, 20XX
  </div>
 </div>
</div>
```

Step 2: Overwrite the Grid of the Desktop Display and Layout with the New Mobile CSS File

The next step after creating the HTML file is to create two cascading style sheets: one that will be used to style your regular website and one that will be called upon when a mobile device is detected. Open up the screen.css file in your text editor. You should see the following style declarations:

CSS (styles from the original desktop CSS file)

```
#container
{
  margin: 0 auto;
```

```
  width: 100%;
  background: #fff;
font-family: helvetica,arial,sans-serif;
}

#header
{
  background: #ccc;
  padding: 20px;
}

#header h1 { margin: 0; }

#navigation
{
  float: left;
  width: 100%;
  background: #333;
}

#navigation ul
{
  margin: 0;
  padding: 0;
}

#navigation ul li
{
  list-style-type: none;
  display: inline;
}

#navigation li a
{
  display: block;
  float: left;
  padding: 5px 10px;
  color: #fff;
  text-decoration: none;
  border-right: 1px solid #fff;
}

#navigation li a:hover { background: #383; }

#content-container
{
  float: left;
  width: 100%;
```

```
  background: #FFF url(layout-two-liquid-
background.gif) repeat-y 68% 0;
}

#content
{
  clear: left;
  float: left;
  width: 60%;
  padding: 20px 0;
  margin: 0 0 0 4%;
  display: inline;
}

#content h2 { color: #039; margin: 0; }

#aside
{
  float: right;
  width: 26%;
  padding: 20px 0;
  margin: 0 3% 0 0;
  display: inline;
}

#aside h3 { color: #039; margin: 0; }

#footer
{
  clear: left;
  background: #ccc;
  text-align: right;
  padding: 20px;
  height: 1%;
}
```

Verify that the style declarations are there, and then save screen.css. Next, open up a new blank file in your text editor. Save the blank file as a new CSS file named "mobile-screen.css." This new CSS file is where our primary task will start as we cancel and rework any nonmobile-friendly CSS rules. Let's take a closer look at the style rules from screen.css to decide where we need to apply changes to mobile-screen .css. In looking at the CSS styles from screen.css, we will learn how to remove and edit styles to make our site display for mobile environments. Remember, our goal is to stack the HTML elements on top of each other as we move away from the grid layout to our linear layout.

float: and clear: are style rules that are used to position HTML elements next to each other in columns and grids. #navigation, #navigation li a, #content-container, #content, #aside, and #footer all have grid layout styles applied to them that need to be removed. Add the following styles to mobile-screen.css:

```
#navigation {background:#fff;float:none;}
#navigation li a {display:inline;float:none;}
#content-container {display:block;float:none;}
#content {display:block;float:none;}
#aside {display:block;float:none;}
#footer {display:block;float:none;}
* {float:none;}
html, body {font:12px/15px sans-serif;background:#fff;
  padding:0;color:#000;margin:0;}
```

(Note: The full source code is available as a download if you would like to cut and paste. Follow the download link at the beginning of this project.)

A quick word about these changes: In addition to our specific elements, we added some global rules that will ensure some consistency across mobile platforms. The * {float:none} and html, body rules remove all floats we may have missed and zero out basic styles setting the font size, text, color, background color, and margins for our mobile display. Save these new CSS rules into mobile-screen.css, which we created earlier, and place it in the same directory as screen.css. With the creation of mobile-screen.css, we have now overridden the major desktop styles from screen.css, but there is more work to do as we customize for mobile.

Step 3: Create a Third CSS File to Combine Previous CSS Files and Take Advantage of Cascading Rules

Now that we have two sets of styles—one desktop style (screen.css) and one mobile style that will override the desktop style when a mobile device is detected (mobile-screen.css)—we are going to create a third CSS file that will refer to both screen.css and mobile-screen.css. Our third CSS file will contain the references to the two files and use the cascading rule (i.e., style sheets links later in the document will override earlier CSS rules) to make our site adapt to different devices. Open up your text editor and add the following three lines, and save the file as "core.css" in the same directory as screen.css and mobile-screen.css:

```
@import url("screen.css");
@import url("mobile-screen.css") handheld;
@import url("mobile.css") only screen and (max-device-
  width:480px);
```

CSS works by applying style rules by what comes last in a reference. We rely on this fact to allow the styles we call later in mobile-screen.css to override the previous styles from our screen.css file. We also direct when to apply the mobile-screen.css by assigning the handheld device declaration and some CSS logic based on the size of the screen (i.e., max-device-width:480px) to catch all mobile browsers and let them know which styles to use and which to ignore. The logic I mention is a powerful new function of CSS3 named "media queries," which allows us to serve different styles based on the size of a browser window or the orientation of a device. After you have saved core.css in the same directory as the other CSS files, move on to the next step.

Step 4: Finalize and Customize the Styles for the Mobile View (with mobile.css)

To customize our mobile view, we need to add our final styles to our last (fourth) CSS file, which we will name "mobile.css." We made a reference to this mobile.css file in the last line of our core.css file, but we had not created it yet. This is our chance to fine-tune the display for our mobile site. We will want to make sure our links have padding for touch interactions. We'll also add some styles that make button-like styles appear for our links, align the text centrally, standardize some padding and widths, and, finally, re-create the global navigation as buttons. Create a new file in your text editor and add the following lines:

```
body {width:100%;font:100%/1.4 helvetica,arial,sans-
  serif;-webkit-text-size-adjust:100%;}
a {padding:45px;border-radius:3px;background-color:
  grey;}
#header > h1 {text-align:center;}
#navigation {text-align:center;}
#navigation ul {list-style-type:none;margin:5px auto 0
  auto;}
#navigation ul li {display:inline;padding:0 5px;}
#navigation ul li a {padding:3px 10px;text-decoration:
  none;-moz-border-radius:3px;-webkit-border-radius:
  3px;border-radius:3px;}
#navigation ul li:nth-child(1) {padding-left:0;}
#content-container {width:100%;background-image:none;}
```

```
#content {width:100%;margin-left:2%;}
#aside {width:100%;margin-left:2%;padding-top:10px;}
#footer {text-align:center;}
```

(Note: The full source code is available as a download if you would like to cut and paste. Follow the download link at the beginning of this project.)

This is really just a start, and you can tweak until you are satisfied. When you are done, save the file as "mobile.css" in the same directory as the others.

Step 5: Bring All of the CSS Files into the HTML Page

Our last step is to add file links for all four of our CSS files into our HTML document. Notice that we have not changed any of the HTML markup for the file with the exception of the <meta viewport tag. What we have changed is the styles for how to display and lay out each of the HTML elements. Within our HTML documents we will add these lines to the <head> of the document:

```
<link rel="stylesheet" href="core.css" media="screen"/>
<link rel="stylesheet" href="mobile.css" media=
 "handheld, only screen and (max-device-width:480px)"/>
```

With these lines, each of our styles will cascade into the HTML document according to the order and logic we assigned. We also added some additional media queries to direct our mobile browser when to apply the mobile.css file. Note how mobile.css is referenced directly in the HTML document and again in the core.css file. We make this double reference to accommodate older browsers that will not recognize media queries. Finally, save the HTML file as "index.html," and load it up in both your desktop and mobile browser to see how it adapts to the different browsing environments.

▶ BUILD A MOBILE SITE USING JAVASCRIPT FRAMEWORKS (JQUERY MOBILE)

View the demo here: http://www.alatechsource.org/techset/; filename = touch-jquery

Download the files here: http://www.alatechsource.org/techset/; filename = touch-jquery.zip

JavaScript frameworks (jQTouch, JQuery Mobile, iUI) give us a chance to apply some of our standard web development skills to build mobile sites. One of the other primary advantages of these frameworks is the way they mimic the display and design interactions of native mobile apps. For example, a mobile website built using jQuery Mobile feels like an app that you might download from an app store and install on your mobile device. You get the familiar look and feel, the quick loading of data, and common navigation design and cues that are part of the native platform without having to build the app using proprietary or new programming languages. Additionally, the frameworks have been tested across multiple mobile environments, making for a compliant solution that works across the major mobile platforms, including iOS (iPhone) and Android. In this project, we will look at how to use jQuery Mobile to build a mobile site that performs like an app on the major mobile platforms. Our finished project will look like Figure 5.15.

Step 1: Create the index.html Page and Set Up the jQuery Mobile Framework

We start by downloading the project .zip file (http://www.lib.montana.edu/~jason/files/touch-jquery.zip), unzipping the .zip file, and opening the index.html page in a text editor like Dreamweaver or Notepad. jQuery Mobile uses a single HTML page for all of its markup. This means that all of our edits to create our different page views for the app will be contained in a single HTML file. There is no need to create separate HTML files for each page. This first step

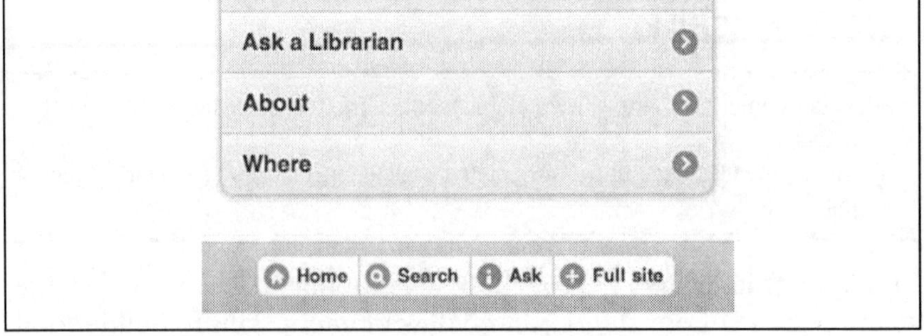

▶ Figure 5.15: View of jQuery Mobile Template for Library Websites

involves calling the jQuery Mobile framework into the index.html page. The framework consists of a set of CSS styles (jquery.mobile-1.0a3.min.css) and two JavaScript files (jquery-1.5.min.js and jquery.mobile-1.0a3.min.js). For the project we are going to use the jQuery content delivery network to link to the JavaScript and CSS files that will make the site go. Here are the relevant lines from index.html:

```
<link rel="stylesheet" href="http://code.jquery.com/
 mobile/1.0/jquery.mobile-1.0.min.css">
```

The preceding link markup brings in the CSS styles that provide instructions for how to display the site in a mobile setting. The following two script tags bring in the jQuery basic framework JavaScript file and its mobile counterpart:

```
<script src="http://code.jquery.com/jquery-1.6.4.min
 .js"></script>
<script src="http://code.jquery.com/mobile/1.0/jquery
 .mobile-1.0.min.js"></script>
```

These two scripts provide the behavior for the app and give us access to special utility functions based on the HTML markup we provide. They are two simple lines, but they create the core functions for the app.

In the jQuery Mobile framework, the HTML markup is how we communicate to the JavaScript what to do with the elements we code into the HTML page. The bulk of our time in building the app is spent in creating this specific markup for the single page, index.html. First, we need to create our <html>, <head>, <title>, and <meta> tags for the page. These tags will help identify the type of file we are serving, give some display directives, and name the file. Look at index.html in your text editor, and verify that the following lines are there:

```
<!DOCTYPE html>
<html>
<head>
<meta http-equiv="Content-Type" content="text/html;
 charset=UTF-8">
<meta name="viewport" content="width=device-width,
 minimum-scale=1, maximum-scale=1">
<title>Mobile @ Your Library</title>
<link rel="stylesheet" href="http://code.jquery.com/
 mobile/1.0/jquery.mobile-1.0a3.min.css">
<link rel="stylesheet" href="http://www.lib.montana
 .edu/~jason/files/touch-jquery/meta/styles/m.css">
```

```
<script src="http://code.jquery.com/jquery-1.6.4.min
 .js"></script>
<script src="http://code.jquery.com/mobile/1.0/jquery
 .mobile-1.0.min.js"></script>
```

We open the HTML file with the !DOCTYPE declaration and <html> tag to let the browser know that we will be using HTML markup for the file. Next, we bring in our <meta> tags to add information about the character set and type of file we are serving. Notice the <meta name="viewport"> tag, which is a mobile-specific tag that tells the web browser to adapt the HTML content to the width of the mobile device and sets the scale of the display according to the scale of the mobile device. Finally, we add the <title> tag to give the HTML document (and our app) a title. This is the title you will see displayed in your web browser's chrome or tab.

Note: I included the markup for the CSS and JavaScript files mentioned in Step 1. These files are brought in the <head> tags of the HTML file, and I wanted to show their position in the context of the whole HTML document.

Step 2: Create the HTML Markup Used for Layout and Display of the Homepage

Once we move out of the <head> HTML markup, we get to the skeleton for the app. This markup will provide the cues for our scripts to act upon and create the different page views for the app itself. First, let's build a homepage for the app. We begin by closing the </head> tag and opening the <body> tag

```
</head>
<body class="ui-mobile-viewport">
```

Next, we create our homepage view. Each page view in a jQuery Mobile app will have a <div> with a data role of header, content, or footer depending on the type of data you want the <div> to contain. Here's the complete page markup:

```
<!-- Start of home page -->
<div data-role="page" id="home" data-theme="d">

<div data-role="header">
  <h1>Mobile @ Your Library</h1>
</div><!-- /header -->

<div data-role="content">
<ul data-role="listview" data-inset="true">
```

```
    <li><a href="#search">Search</a></li>
    <li><a href="#hours">Hours</a></li>
    <li><a href="#ask">Ask a Librarian</a></li>
    <li><a href="#about">About</a></li>
    <li><a href="#where" data-rel="dialog" data-
     transition="pop">Where</a></li>
  </ul>
  </div><!-- /content -->

  <div data-role="footer" data-id="myfooter" data-
   position="fixed">
    <div class="controls" data-role="controlgroup" data-
     type="horizontal">
    <a href="#home" data-role="button" data-icon="home">
     Home</a>
    <a href="#search" data-role="button" data-icon=
     "search">Search</a>
    <a href="#ask" data-role="button" data-icon="info">
     Ask</a>
    <a href="/index.php" rel="external" data-role=
     "button" data-icon="plus">Full site</a>
    </div>
  </div><!-- /footer -->

  </div><!-- /page -->
```

By assigning the data-role="page" to the first <div> in our markup we are letting the jQuery Mobile script know that we are creating a unique page. In addition, our markup id="home" data-theme="d" identifies this particular page as the homepage for our app and brings in the gray gradient theme that creates the look and feel for the page.

The homepage for our app has our primary navigation with links to all the other page views. After we create the <div data-role ="header"> and add our page title in an <h1> tag, we build the actual content that will allow us to move through the application. The markup <div data-role="content"> lets the jQuery Mobile script know that we are into the actual unique content for the page. Our homepage content is essentially a list of links in a <ul> container. Here's the relevant markup:

```
  <ul data-role="listview" data-inset="true">
    <li><a href="#search">Search</a></li>
    <li><a href="#hours">Hours</a></li>
    <li><a href="#ask">Ask a Librarian</a></li>
```

```
    <li><a href="#about">About</a></li>
    <li><a href="#where" data-rel="dialog" data-transition=
     "pop">Where</a></li>
  </ul>
```

By giving the <ul> the data-role="listview" data-inset="true" properties, we use the script's functions to show a mobile list style with a margin around the list. All links for other page views are internal links using the # and the id for each unique page.

We close the page with our global footer, which contains a data-role ="controlgroup" where our global controls for the app will be. In this case, it is a list of links to certain pages. We are able to use some default button appearance and icons on our links by assigning data-role="button" and data-icon= to each link <a href>. Note how we included an external link to the full desktop version of the site using the markup <a href="/index.php" rel="external" data-role= "button" data-icon="plus">Full site</a>. Here is the complete footer markup:

```
<div data-role="footer" data-id="myfooter" data-
 position="fixed">
  <div class="controls" data-role="controlgroup"
   data-type="horizontal">
  <a href="#home" data-role="button" data-icon="home">
   Home</a>
  <a href="#search" data-role="button" data-icon=
   "search">Search</a>
  <a href="#ask" data-role="button" data-icon="info">
   Ask</a>
  <a href="/index.php" rel="external" data-role=
   "button" data-icon="plus">Full site</a>
  </div>
</div><!-- /footer -->
```

Step 4: Create the HTML Markup Used for Layout and Display of the Search Page

We have five page views for the app, and each of the page views has a global header and footer that is the same for all pages. Moving forward, we'll look at the markup that creates the unique page content. Our search page is the first example. Here's the complete page markup:

```
<!-- Start of search page -->
<div data-role="page" id="search" data-theme="d">
```

```
<div data-role="header">
  <h1>Mobile @ Your Library</h1>
</div><!-- /header -->

<div data-role="content">
  <h2>Search the Library</h2>
    <form action="forms-results.php" method="get">
    <fieldset>
    <input type="text" value="" />
    <div data-role="fieldcontain">
    <label for="select-options" class="select">Choose
     an option:</label>
  <select name="select-options" id="select-options">
  <option value="option1">Books, Videos, &amp;
   More</option>
  <option value="option2">Articles</option>
  <option value="option2">Databases</option>
  </select>
  </div>
  <button type="submit">Find</button>
  </fieldset>
  </form>
  <p><a href="#home" data-direction="reverse">Back to
   home</a></p>
</div><!-- /content -->

<div data-role="footer" data-id="myfooter" data-
 position="fixed">
  <div class="controls" data-role="controlgroup"
   data-type="horizontal">
  <a href="#home" data-role="button" data-icon="home">
   Home</a>
  <a href="#search" data-role="button" data-icon=
   "search">Search</a>
  <a href="#ask" data-role="button" data-icon="info">
   Ask</a>
  <a href="/index.php" rel="external" data-role=
   "button" data-icon="plus">Full site</a>
    </div>
</div><!-- /footer -->

</div><!-- /page -->
```

Our first move with any page view in jQuery Mobile is to uniquely identify the page. The markup <div data-role="page" id="search" creates this unique page cue that tells the JavaScript where to create a

page view. Our search page has an input form, and we are able to leverage some of the custom styles and interaction patterns that are part of the jQuery Mobile JavaScript. The search form markup is within the <div data-role="content"> inside the </form> tag. (Note: To make the form work with your library, set the action= to point at your preferred library collection search or your library catalog search URL. Look for the <form action="forms-results.php" method= "get"> HTML tag in the source code to make this change.) jQuery Mobile will automatically scale the <input type="text" value="" /> tag to fit within mobile device dimensions, but to take advantage of the touch button stylings we need to wrap our form options in a special div tag. The markup looks like this:

```
<div data-role="fieldcontain">
  <label for="select-options" class="select">Choose an
   option:</label>
  <select name="select-options" id="select-options">
  <option value="option1">Books, Videos, &amp; More
   </option>
  <option value="option2">Articles</option>
  <option value="option2">Databases</option>
  </select>
</div>
```

By giving data-role="fieldcontain" to the <div>, we are telling the JavaScript to style our options as buttons that are optimized for touch interaction. (Note: To use all of the options with your search form, you will need to add the correct values or parameters needed to direct your local search to perform a specific search action.) The final notable markup is our creation of a dynamic "back" button for the page. Here's the relevant markup:

```
<a href="#home" data-direction="reverse">Back to home</a>
```

It is a simple <a> tag with a link to our home page view (href= "#home") and a custom attribute data-direction="reverse" that notifies the JavaScript to move the current page view back to our homepage.

Step 5: Create the HTML Markup Used for Layout and Display for the About, Hours, and Ask Pages

The About, Hours, and Ask a Librarian page views are identical in markup. The only difference is the actual text or copy on the page. We'll work through the Ask a Librarian page view to show how to build these page views. Here is the full Ask a Librarian page markup:

```
<!-- Start of ask page -->
<div data-role="page" id="ask" data-theme="d">

  <div data-role="header">
    <h1>Mobile @ Your Library</h1>
  </div><!-- /header -->

  <div data-role="content">
    <h2>Ask a Librarian</h2>
    <p><strong>Text Us!</strong></p>
    <p>Send us a question as a text message at 1-406-
     219-1060. See <a href="http://www.lib.montana
     .edu/ask/textalibrarian.php" rel="external">more
     info and the demo</a></p>
    <p><strong>Email Us!</strong></p>
    <p>Send a <a href="http://www.lib.montana.edu//
     forms/email.php" rel="external">question via our
     email form</a>.</p>
    <p><strong>Call Us!</strong></p>
    <p>Voice Phone: 1-406-994-3171 (leave a message if
     we are not there)</p>
    <p><a href="#home" data-rel="back">Back to home
     </a></p>
  </div><!-- /content -->

  <div data-role="footer" data-id="myfooter" data-
   position="fixed">
    <div data-role="controlgroup" data-type="horizontal">
    <a href="#home" data-role="button" data-icon="home">
     Home</a>
    <a href="#search" data-role="button" data-icon=
     "search">Search</a>
    <a href="#ask" data-role="button" data-icon="info">
     Ask</a>
    <a href="/index.php" rel="external" data-role=
     "button" data-icon="plus">Full site</a>
    </div>
  </div><!-- /footer -->

</div><!-- /page -->
```

We begin again by uniquely identifying the page. The markup <div data-role="page" id="ask"> creates our unique page view. After that it is just a matter of adding the markup to the <div data-role="content"> section to display the unique copy for the page view. The unique copy is just a series of <h2>, <p>, and <a> tags. One important note is the additional rel="external" attribute in the <a> tag. Any time you want to leave the jQuery Mobile environment, you need to give a cue to the JavaScript. All links to external resources need to follow this convention to turn off the jQuery Mobile framework, as shown here:

```
<a href="http://www.lib.montana.edu/ask/textalibrarian
 .php" rel="external">
```

The final part of this step is to carry out all of the same markup routines on the About and Hours page views.

Step 6: Create the HTML Markup Used for Layout and Display for the Where Page

The Where page view breaks from our previous markup routines for a couple of reasons. First, we are launching an internal viewer using the dialog popover effect. And second, we are bringing in some dynamic data from the Google Static Maps API to create a map image. To tell the JavaScript to overlay the map on top of the homepage, we create an HTML markup cue into the dialog function to set the link in the global navigation:

```
<a href="#where" data-rel="dialog" data-transition=
 "pop">Where</a>
```

With this markup, we are taking advantage of the jQuery Mobile function that will create the overlay effect using the "dialog" call and assign a stylized transition to the popover that loads the map into view. The full markup for the page is actually pretty straightforward:

```
<!-- Start of where page -->
<div data-role="page" id="where" data-theme="d">

  <div data-role="header" data-theme="d">
    <h1>Mobile @ Your Library</h1>
  </div><!-- /header -->

  <div data-role="content" data-theme="d">
    <h2>Where is MSU Libraries?</h2>
    <p><img alt="map to library" src="http://maps
     .google.com/maps/api/staticmap?center=Bozeman,MT&
     zoom=13&size=310x310&markers=color:blue|45.666671,
     -111.04859&mobile=true&sensor=false" /></p>
    <p><a href="#home" data-rel="back">Back to
     home</a></p>
  </div><!-- /content -->
```

```
<div data-role="footer" data-id="myfooter" data-
 position="fixed" data-theme="d">
  <div data-role="controlgroup" data-type="horizontal">
  <a href="#home" data-role="button" data-icon="home">
   Home</a>
  <a href="#search" data-role="button" data-icon=
   "search">Search</a>
  <a href="#ask" data-role="button" data-icon="info">
   Ask</a>
  <a href="/index.php" rel="external" data-role=
   "button" data-icon="plus">Full site</a>
  </div>
</div><!-- /footer -->

</div><!-- /page -->
```

The tricky part is customizing the map for your location. As before, we start by assigning the unique page view value using the <div data-role="page" id="where" data-theme="d"> markup. Next we move on to create the custom map image. The markup for the image should be familiar:

```
<img alt="map to library" src="http://maps.google.com/
 maps/api/staticmap?center=Bozeman,MT&zoom=13&size=310x
 310&markers=color:blue|45.666671,-111.04859&mobile=
 true&sensor=false" />
```

It is a simple <img> tag with a call linking into the Google Static Maps API. To assign a specific location we need to pass a few parameters to the Google Static Maps API. Included among the customizations are:

1. a central location for the map,
2. a zoom level for the map,
3. a size for our map image,
4. a marker color and position using latitude and longitude, and
5. set the map to display for mobile settings without a sensor.

The necessary pieces to enter are a central location and your specific latitude and longitude. Enter your hometown for the central location in the format of {city, state or principality}. To get your latitude and longitude, visit http://gmaps-samples-v3.googlecode.com/svn/trunk/geocoder/getlatlng.html and type in your library address.

Copy the latitude and longitude values that are returned. Enter these position values as a comma separated string {latitude,longitude}. Here's the image tag with cues for where to add your values:

```
http://maps.google.com/maps/api/staticmap?center={YOUR
  HOMETOWN}&zoom=13&size=310x310&markers=color:blue|
  {YOUR LATITUDE/LONGITUDE}&mobile=true&sensor=false
```

Once you have the link filled out, add it to your image <img> tag and you will have a local map to your library location.

Step 7: Assign Styles for Display with a Custom CSS File

After we have created all of the custom page views and markup, you might still want to tweak some of the colors, layouts, and styles associated with your app. In this case, we can override some of the jQuery Mobile default styles with our own style sheet. Our only customization initially will be to center the controls in our global footer with the following CSS styles:

```
/*custom styles for jquery mobile*/
.controls {text-align: center;}
.controls * {margin: 0 auto;}
```

This is a simple change to provide some symmetry to the app, but you could go much further if you wanted to set a different background on the header or style the <h1> and <p> tags with a different font. If you want to customize further, you can add additional style rules to the custom CSS file located at /meta/styles/m.css.

Step 8: Rename the /touch-jquery/ Directory, and Deploy the Mobile Website

The final step is to rename the /touch-jquery/ directory to a mobile-friendly name or convention. Until now, we have been using the /touch-jquery/ directory for development purposes, but with our development done we want to make it publicly addressable in a URL. Rename the /touch-jquery/ directory to /m/ (or whatever makes sense for your library mobile site), and upload or FTP the directory to your live public website. With that change, you can use the /m/ directory as a link from your library homepage to a mobile version of your site. And with these final steps, you have completed your first jQuery Mobile app.

6

MARKETING

- Market Using Traditional Print Media
- Market via the Web
- Optimize Apps or Sites for "Findability" Using Search Engine Optimization
- Market within Your Organization

With the work complete on your mobile app or mobile website, it is time to turn your thoughts toward how to get the word out. And getting the word out is just one of the goals here; you also want to get people using your apps and sites as much as possible. Much like the moving target of mobile application and site development, promotion and marketing of mobile apps and sites is part of an evolving networked ecosystem. This being said, we will look at three primary networks and communication modes that you can use to let your users know about all of your hard work.

MARKET USING TRADITIONAL PRINT MEDIA

Our first method of promotion and marketing involves the familiar modes commonly associated with traditional print media: newsletters, newspaper or local press, bulletin board flyers, and so forth. Here are a few options to get you started:

1. **Write an article for your library newsletter:** Make sure to spread the word of your mobile presence in your library newsletter. A simple blurb detailing what you have created, why people might want to visit it, and a fact or two about how mobile visits are growing will help create a picture of your efforts.
2. **Contact and use your local press:** Local press, independent from your organization, is another valuable communication

source. Contact a local reporter and pitch an idea about how the library is responding to the changing nature of information networks and emphasize your mobile project as one example of this response. Promote your mobile project in the local community events and happenings newspaper pages right alongside your announcements about summer hours.

3. **Conduct a physical flyer campaign:** Don't overlook the simple flyer campaign. Find the community billboards at your local co-op, university student union, or even your library billboard. Place a simple flyer with a library logo, a mobile icon, and a link to your mobile URL.

These are just a few of the possibilities and traditional physical formats for promotion and marketing that can still be very effective. Use your imagination.

▶ MARKET VIA THE WEB

A second method of promotion and marketing focuses on the electronic forms associated with the rise of the web: webpages, blogs, social networks (Twitter, Facebook), and QR codes. In many ways, these online modes will be your primary form of communication about your mobile efforts:

1. **Create an HTML landing page for your app or service:** Build an HTML landing page for your app or service answering the what, why, and how of your mobile project much like the print blurb you created earlier. To get some ideas for app landing pages, check out services like App.net (http://app.net/) or Limelight (http://www.limelightapp.com/), which can help you generate your own landing pages with custom templates and analytics.
2. **Announce your app or service on your library blog:** A blog post announcing your new mobile app or service is another essential promotion and marketing tool. People like to have a point of reference, and in the web world a URL is the lingua franca. People can share a blog post or link to your landing page, and your message will be amplified.
3. **Post about your app or service on social networking sites:** If you have a social networking presence, consider posting about your mobile project. Twitter has a 140-character limit that can help you streamline your message and teach you how to present only the necessary details about your mobile effort. Facebook has an

engaged, yet broad, user base where you can reach many and become part of the "like" and sharing ecosystem. In either case, your social network post should mirror the blurb from your webpage or blog post giving the details of your mobile project and where to go to use it.

4. **Create a Quick Response (QR) code:** QR codes are two-dimensional bar codes that allow dedicated bar-code readers and smartphone cameras to read and draw information from the bar code itself. QR codes, by their nature, are tied to mobile devices and offer a perfect promotion medium in the mobile setting. In the promotion of your mobile project, a QR code could be created to point to the URL for your mobile site or a mobile services information page. A user could scan the QR code and be directed to these sources.

QR Code Resources

Creating QR codes is becoming easier and easier.

See the Kaywa QR code generator at http://qrcode.kaywa.com/ or the Google Chart API wizard at http://code.google.com/apis/chart/image/docs/chart_wizard.html to create your own.

To learn more about creating QR code campaigns, check out Joe Murphy's *Location-Aware Services and QR Codes for Libraries* (THE TECH SET #13).

The possibilities for promotion and marketing using electronic forms continue to evolve, but these options can get you started in creating a digital presence for your mobile apps and services.

▶ OPTIMIZE APPS OR SITES FOR "FINDABILITY" USING SEARCH ENGINE OPTIMIZATION

A third method of promotion and marketing involves the established practice of search engine optimization (SEO) for making our apps and sites discoverable in search engines. Among the methods of promotion and marketing discussed in this chapter, the SEO method ranks as one of the most important. SEO is the practice of assigning keywords, descriptions, and other metadata to your webpages and building strategic links into your content to allow for indexing by web search engines like Google and Yahoo!. When done right, SEO will help make your app or site more "findable" by making your resources more discoverable (i.e., highly ranked) in search engine result pages.

SEO is interesting in practice, because it bridges two worlds: the end user searching and the search engine robots that crawl web content and add it to search engine indices. (Note: We are focusing on a mobile SEO strategy, but the lessons and procedures detailed here can also be applied to your full website.)

With these dual users (humans and robots) in mind, our SEO strategy for promoting our mobile app or service will follow two steps:

1. Placing keyword and description markup within your mobile site, landing page, or mobile app
2. Creating a sitemap.xml file for your mobile site and/or mobile landing page

These steps are a good place to begin with SEO and represent some essential moves we can make to help promote and market our digital files, but this is just a start.

SEO Resources

To learn more about the complex and varied field of SEO, have a look at Google's Search Engine Optimization Starter Guide at http://goo.gl/h4DY5.

Step 1 begins with keyword research for our mobile project. We will be defining the key terms that people will use to search for your project and communicating those terms to the web indexing robots. To get started, look closely at the narratives you created when drafting the "why, what, and how" passages for your blog post, landing page, or any of the earlier promotional steps above. Pull out all the key terms you can find. Once you have your key terms, run them through the Google AdWords Keyword Tool (http://goo.gl/fDjO4) one at a time. The Keyword Tool will return suggestions related to your term. Keep a list of all the terms you generate. Once you are satisfied that you have a list of useful terms, add them to your HTML files or blog post related to the mobile project. Web indexing robots rely on certain HTML tags to help guide them in defining the "aboutness" of a page. In this SEO example, we are going to work with our landing page HTML file to demonstrate. The same practice could be done with your blog or social network posts in revising the post and post title to incorporate the key terms you discovered. If you have created an HTML landing page, open up the file and add your terms and variations on the terms into the following primary HTML tags. Pay particular attention to the instructive comments inside the HTML tags:

```
<title>REVISED TITLE USING PRIMARY KEY TERMS</title>
<meta name="description" content="YOUR DESCRIPTON
 REVISED AND LOADED WITH KEY TERMS" />
<meta name="keywords" content="YOUR FULL LIST OF KEY
 TERMS" />
<h1>FIRST HEADER ON PAGE REVISED TO USE PRIMARY KEY
 TERMS</h1>
```

You can also insert the revised passage you entered into your <meta name="description"> tag into the main copy of your page, probably in some <p> tags that are part of the main HTML landing page markup. When finished, save the HTML file. By adding all of this "keyword density" into the HTML page, you are giving the web indexing robots essential classification information and enriching the search term possibilities for anybody who might use a search engine to find your mobile project.

A related subset of this SEO mode would be optimizing your app or site for the specific app store and mobile site marketplaces that are becoming a primary mode for discovering new mobile resources. We can reuse the research we did to find keywords and descriptions in the first step and make sure to add these terms to the title and description for our mobile app. Add these new terms by logging into your account for the particular application marketplace you are interested in updating and adding the new terms and revisions to your app profile. In this case, we are looking to improve the search and retrieval for our app in the Android App Marketplace or Apple App Store (depending on the type of mobile app we are marketing), not Google or Yahoo!. App stores and other mobile site marketplaces have different search algorithms, so it can be useful to conduct some additional research. A helpful practice here would be to run a few searches in the App Store or Android Marketplace to determine what other related library mobile apps are using in their keywords and descriptions. If you find some unique or useful terms, you can add them to the metadata (title, description, keywords) for your app.

Findability Resources

To learn more about improving mobile app discovery, take a look at "Making Your Mobile App More Discoverable" from Mashable.com (http://mashable.com/2011/06/24/improve-mobile-app-discovery/).

The second step in our SEO strategy involves letting the web indexing robots know which files to use when building a search index for the mobile files that are part of our site or are related to our mobile app. We are going to use the sitemap protocol (http://www.sitemaps.org/) to create a sitemap file. What is a sitemap file? From the Sitemaps.org site:

> Sitemaps are an easy way for webmasters to inform search engines about pages on their sites that are available for crawling. In its simplest form, a Sitemap is an XML file that lists URLs for a site along with additional metadata about each URL (when it was last updated, how often it usually changes, and how important it is, relative to other URLs in the site) so that search engines can more intelligently crawl the site.

In our case, we are going to create a sitemap file that lists the files associated with our mobile apps and services. For more information about mobile sitemaps, visit http://goo.gl/VCVxY. If you are just interested in getting started, open up a text editor and add the following lines:

```
<?xml version="1.0" encoding="UTF-8" ?>
<urlset xmlns="http://www.sitemaps.org/schemas/sitemap/
 0.9" xmlns:mobile="http://www.google.com/schemas/
 sitemap-mobile/1.0">
```

These lines identify the file as an XML file and tell our web indexing robots that this XML file follows the sitemap protocol. Next, we add the list of URLs that need to be associated with our mobile apps and services. In your text editor after the previous, enter all of the URLs you want to list using the following XML. Each file will get its own <url> tag, and the value of the URL will be placed in the <loc> tag:

```
<url>
<loc>http://mobile.example.com/article100.html</loc>
<mobile:mobile/>
</url>
```

When you are finished adding all of your files, add the closing tag for the sitemap XML file:

```
</urlset>
```

Once the file is complete, save the file as "sitemap.xml" in the directory on your website where your mobile project files or mobile HTML landing pages are located. Once your mobile sitemap.xml file has a home, the last step is to let search engines know that the file exists. To submit the full URL for your sitemap file to Yahoo! and Google, visit the following URLs and follow the instructions (other search engines have different resources for sitemap submission, but checking their documentation can guide you in their sitemap submission routines):

- Yahoo Mobile Sitemaps: http://siteexplorer.search.yahoo.com/mobilesubmit
- Google Sitemaps: http://goo.gl/bvgNM

And, with that, you have finished the SEO method for promotion and marketing of your mobile apps or services.

A promotion and marketing strategy that uses the modes described will help let your users know about your mobile presence. And these modes are not mutually exclusive. In fact, combining modes may create some of the most effective messages for your users. Imagine a flyer on a community bulletin board with the message "Scan the code below to enter another world" with an accompanying QR code that links to your mobile catalog or to a children's storytime schedule optimized for mobile browsing. It really is up to you.

▶ MARKET WITHIN YOUR ORGANIZATION

To this point, we have focused on external promotion and marketing, but there is another audience that we need to consider: our internal staff. Internal marketing has a different focus and is not as much about making things findable as it is about empowering staff to carry the message about your mobile projects and promote your mobile tools to patrons. Here are some simple methods for getting the word out to your internal library staff regarding your mobile projects or services:

1. **Send an internal e-mail to all library staff:** The preferred medium for communicating will depend on your organizational culture, but the important thing to consider is what format works best to reach all people within your organization. E-mail works well, as it is typically a required form of communication in modern business settings. Send an e-mail to all library staff announcing your mobile app or service. Be sure to include the URL for the app or service as well as the what, why, and how.
2. **Schedule a public presentation and demo of the mobile app or service:** Offer a face-to-face presentation and demo of the app or service. Or invite all library staff to a brown-bag discussion of why the mobile project exists and show how it works. People can engage more directly in these settings, and their understanding of your project will only improve.

3. **Create a brochure or business card for all public service staff:** Create a handout with simple talking points and the URL for the app or service for all library staff. In addition, consider a small brochure or business card for public services staff to have on hand at the circulation or reference desk explaining the mobile app or service.

The idea is to enable advocacy, and these methods can help reinforce the reasoning behind your mobile app or service and allow other library staff to carry your message.

7

BEST PRACTICES

- Consider the Simplicity of the Mobile Platform
- Consider the Basics for Mobile App Design and Development
- Consider Design Options

CONSIDER THE SIMPLICITY OF THE MOBILE PLATFORM

When looking to define best practices for a transitional platform like mobile applications, it can be useful to consider what makes mobile development and design different from desktop design and development. There is simplicity to the mobile platform that lends itself to minimal design and limited app features. This austerity is echoed in the following quote from the iPhone interface guidelines:

> In iPhone apps, the main function should be immediately apparent. Minimize the number of controls from which users have to choose. (iPhone User Experience Guidelines, http://goo.gl/4JZUK)

With mobile, you have a different set of users: people concerned with locations, status checkers, quick seekers of info, people monitoring updated info (news or stocks), and so forth.

Mobile development also has a different set of content requirements: location services, quick lookup services, contact services, and the like. New tactile modes of interaction are also part of the mobile development environment, including touch for click, spin orientations, and drag and pinch. Each of these direct manipulation methods creates different design decisions like the need for whitespace or the dynamic resizing of pages. Finally, you have the technical considerations for mobile browsing, such as screen width, battery power, and connectivity and bandwidth issues. Considering these factors, you can start to see some essential design features: simple navigation, whitespace, room for touch interactions, horizontal scrolling of content, minimized code and images, only essential app functions, and so forth. A closer look at the mobile app for Flickr.com demonstrates how these mobile platform considerations play out in development (see Figure 7.1).

The Flickr.com mobile landing page immediately provides actions for a mobile user: create account, sign in, explore, and search. There are no slide shows or previews of content unless you choose to "Explore." Large buttons enable even the largest fingers to navigate

▶ **Figure 7.1: Mobile View of Flickr.com**

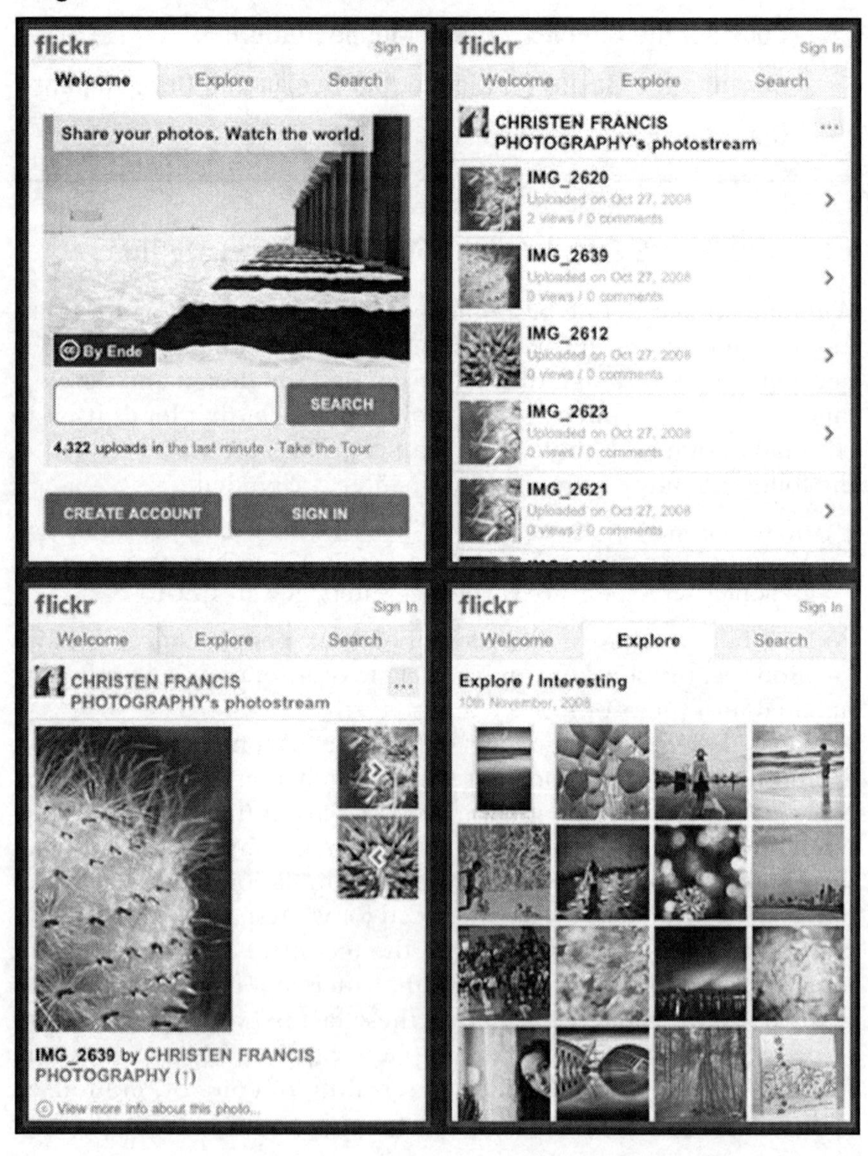

using touch interaction. Thumbnails are small and optimized for the smaller connections and bandwidth available to the mobile browser. The photo browse has a horizontal-scroll single-column layout that allows for unlimited viewing of content within the common widths of the mobile devices. Navigation is pared down to three common user tasks: welcome, explore, and search. Even the item view of an individual photo (see Figure 7.1, bottom left screen) shows an optimized image with enough information to let you know what the image is but not enough to clutter or confuse. It is all very calculated and measured, and good mobile design and development plays within these limits.

▶ CONSIDER THE BASICS FOR MOBILE APP DESIGN AND DEVELOPMENT

As has been mentioned, mobile design and development is different and brings with it new challenges, but it's not all about limits. Mobile design and development considerations can also be liberating. There is something satisfying about being really efficient in your design. Simple as a development aesthetic can open you up to possibilities that you could not envision in the desktop environment with its complexity and large screen real estate. As you start to look at what you might do to design and develop for mobile, here are some foundational best practices and tips to stimulate your "mobile thinking."

Tip 1: It's a Touch Environment, Not a Typing Environment

We are moving away from the indirect manipulation methods of the desktop (think mouse) to the more natural, direct manipulation method of touch. Functions like autocomplete are fundamental to mobile interfaces. Give your users a chance to navigate with their thumb if needed. You can do this by focusing on buttons and links that perform core actions for your app.

Tip 2: Narrow Your Focus

Do a few things really well. Focus on what the main function of your app will be. Mobile design and development doesn't allow for all the bells and whistles of its desktop cousin. This is a good thing. The most successful mobile apps try to solve a single problem. Ask yourself, "What are the core actions my app needs?"

Tip 3: Know Your Audience

Don't do mobile for mobile's sake. Learn about the behavior of your users. Run the analytics, and look for mobile browser types and mobile operating systems in your web stats. Poll your users, asking what they need from a mobile app for your library. Check out PollEvery where.com or SurveyMonkey.com for some simple tools for collecting data. You might be surprised how library patrons are using mobile.

Tip 4: Learn from the Masters

There is always an opportunity to reverse engineer. If you see an app you like, study the code. Apple has released sample apps that you can learn from with their iOS software development kit (SDK) available at developer.apple.com/programs/ios/develop.html. With "View Source" of m.flickr.com in a web browser, you are sure to pick up how the leaders make things go. You can even emulate a mobile browser with the "User Agent Switcher" (Firefox add-on) available at addons .mozilla.org/en-US/firefox/addon/59/.

Tip 5: The Mobile User Interface Is about Direct, Natural Manipulation: Targets, Gestures, and Actions

Think in terms of the new forms of interaction. Tactile navigation creates unique design and development opportunities, and it is an evolving user interface. Consider these standard user interface actions that have emerged in the past couple of years: tap, double tap, drag, flick, pinch, spread, press, press and tap, press and drag, and rotate/spin (gyrometer). Touch as an interface requires large targets. Apple recommends a minimum target size of 29 px wide by 44 px tall. Plan your mobile interface accordingly.

Tip 6: In Mobile Environments, Speed, Performance, and Optimization Are Essential

Consider the typical use case scenarios for your mobile app: people on the move looking for quick information. Respect the need to have quick, unfettered access to your info. You can do this by limiting images, HTML markup, or bloated code. Quick development tips to follow include these:

- Limit HTML pages to 25 KB to allow for caching.
- Limit images to under 1 KB when possible.
- Compile your scripts using the native software development kits.

- "Minify" your scripts and CSS
  - JSLint to minimize JavaScript; http://www.jslint.com/
  - CleanCSS to minimize CSS; http://www.cleancss.com/
- Take advantage of HTML5 and CSS3.

Tip 6: Respect the URL and the App Store Distribution Model

Think about how your app will be distributed and found. Consider a descriptive, intuitive name and proper metadata to make your app "findable" in the cluttered app store marketplace. When thinking in terms of mobile web apps, follow the "m" conventions. See m.delicious .com or lifeonterra.com/m/ as examples of good mobile URLs. If you do need hierarchy for organizing your app, keep categories (directories) short and sweet.

Tip 7: Take Advantage of Mobile Web Browsers' Capabilities When Possible

Mobile web browsers are cutting edge, and new standards like HTML5 are starting to enable native app functionality. The WebKit Browser Engine that runs within Mobile Safari and the Android web browser support many HTML5 and CSS3 features like geolocation, device orientation, offline storage for limited connections, and even media queries to determine device screen dimensions. There will be times when creating a mobile web app with most of the functionality of a native app will make sense. Learn when to make that call.

Tip 8: Allow for User Choice, and Build Interfaces to Accommodate for Failure

Accommodating for failure may seem counterintuitive, but you want people to be able to make mistakes and then recover while using your app. A common method of allowing for mistakes is making sure there is a "back" or "return to home" function on all the views of your app. Another means of allowing for user choice is to have a link to your full desktop website within your mobile app. Giving this control to the user creates a forgiving environment for the unanticipated actions that people might have for your app. If you want to help even further and create a truly responsive app, you might consider "sniffing" for user agents and operating systems and serving different app views for different users. Here's a sample JavaScript detection script looking for iPhone and iPod users (you can do the same for Android and other users):

```
<script type="text/javascript">
if ((navigator.userAgent.indexOf('iPhone') != -1)
||(navigator.userAgent.indexOf('iPod') != -1))
{
document.location = "#";
}
</script>
```

Tip 9: Use Mobile Templates, Code Libraries, and SDKs

At times reinventing the wheel may be necessary, but for simple and quick development, mobile templates, code libraries, and SDKs are a good choice. These development templates and software kits have been vetted and tested across platforms and allow for rapid deployment. Common user interaction tasks are built in and provide a standard interface for typical app functions.

Common Templates and Kits

- Mobile JavaScript code libraries
  - jQTouch; http://jqtouch.com/
  - JQuery Mobile; http://jquerymobile.com/
- Native/mobile web app packages
  - Sencha Touch; http://www.sencha.com/products/
  - PhoneGap; http://phonegap.com/
  - Apple iOS SDK; http://developer.apple.com/devcenter/ios/index.action
  - Android App Inventor; http://info.appinventor.mit.edu/

Tip 10: Know Mobile Design Conventions, and Study the Mobile Interface and Device

To be a good mobile designer and developer, you have to be a good mobile user. This means practicing on and learning to use the mobile platform. You can use mobile emulators to get an idea, but your best course of action is to become a mobile user. This is a call to invest in the technology on which you are building apps. Make the case for an iPad as a department resource to your library director. Purchase an Android phone and recoup some of the costs as a business expense. Be creative. As you become familiar with the mobile platform, your design ideas will grow, and you will start to recognize what works best. Consider the following mobile design conventions:

- The "one-column layout" is the best means to allow for extended views of your data. In this case, scrolling is your ally.

- The limited screen space on mobile devices means whitespace is your friend. Give users a chance to scan your app without being overwhelmed.
- In the same vein, embrace an economy of language. Verbose and intricate labels will detract from the mobile user experience. Keep it simple.
- Design and develop using the "quick visit" architecture. The mobile environment is not about wayfinding and serendipity. Build your app views to allow for quick information and focused, simple tasks.

▶ CONSIDER DESIGN OPTIONS

Incorporate Mobile User Interface Design Patterns

User interface design patterns document good solutions to common interface questions or problems. At times, the design patterns become expectations, and breaking from these common understandings about how an interface should work can lead to poor experiences in using an application. We are in the native stages of mobile design and development, but, even in this beginning stage, it is clear that we are seeing some patterns emerge for design in mobile environments.

Limit to Simple Navigation with the Primary Functions/Tasks

Keep the main navigation to three or four links related to the essential actions for your app. Look at it this way: A mobile app is a series of layers. The main navigation for your app should be basic and let the user choose to move down (or up) into your app hierarchy, views, and actions.

Design for a Single Frame or Single Window

Users experience your application as a collection of screens. Layered and modal applications that allow users to choose what views they need to see are a useful design pattern. Design for linear or horizontal scrolling, as these methods extend the device real estate. Remember: At any given time, an app works in a single frame or window.

Use Popovers and Alerts to Communicate Status of Actions

Because an app can occupy only a single window, mobile developers rely on popovers or alerts to tell users what is happening in the application. When designing, look to create transparent message windows that display over the top of your main application window.

Maintain Visual Consistency with Your Desktop Website or App

Your branding matters. Look at ways to bring your color scheme and logo from your organization into your mobile presence. Maintaining visual cues shows a relationship with your larger web platform. Mobile is a unique extension of your other applications but not a completely distinct entity.

Design to Use Full-Device Widths

Widths vary across mobile devices. Good mobile design adapts to a device by using all of the screen available. Avoid columns and padding that reduce the amount of usable pixels on the (limited) mobile screen. Use all the available width for links, list elements, text inputs, and all possible focusable elements.

Make Native Device Controls the Default Controls

Use native menus, settings, and toolbars when possible. There is no need to reinvent the wheel. User expectations for mobile devices weigh heavily here. If you are going to break from the convention of a top status bar and a bottom settings bar, you must have a compelling reason to do so.

Set Up Actions and Views within a Hierarchy of Priority

Not all layers of an app are created equal. Mobile design requires smart choices about when to display data and which information is essential to a task. Prioritize the views in your app, and use lists to group categories and actions. Make sure to provide the most-used features at the top.

Build Your App Views with Lists or Tables

Small screens need to scroll. They also need to be organized in small grids for more complex information. Mobile designers are applying scrolling lists and simple table formats to allow for reading and browsing in the mobile environment.

There really is no end to design patterns, and the early stage of mobile design suggests different patterns may emerge to solve some of the challenges of application development for mobile devices. I have outlined here a working design pattern library for mobile that introduces the basic considerations and surveys the current working patterns for mobile interfaces. It's a snapshot of conventions to introduce the differences of developing for the mobile platform and encourage "mobile design thinking."

▶ 8

METRICS

- ▶ Determine Metrics
- ▶ Review Providers' Resources

With the launch of our mobile application or service, it is time to think about how we might measure the success of our app. Metrics are the performance measures we can put in place to help us analyze the efficiencies, user behaviors, and benchmarks of our mobile applications or services. Metrics provide the statistical data that we can use to make decisions about future developments of our app. More importantly, metrics can help us make a business case to managers and library administration about the utility of investing in resources for mobile development.

▶ DETERMINE METRICS

In a native application setting, metrics are determined by analytics software packages and the raw usage data collected by operator portals, marketplaces, and app stores. In the case of mobile websites and mobile web applications, we can use traditional web application server logs and software to collect this same raw usage data. In this chapter, we will look at some of the essential metrics you can use to determine the success of your app or service. We will outline common, generic metrics that should be a part of any analytics reports and data you choose to collect and list a few analytics services that you can use to get started.

In each of the examples I describe, I use the common names for the metrics from the Apple App Store and Google Analytics. For example, "Total Downloads" is the name of a metric from the App Store, whereas "Total Visits" is a name of a similar metric from Google Analytics. In most cases, the metrics are roughly equivalent, and I have followed this convention of naming the common metrics (with the App Store name preceding the Google Analytics name) in hopes of showing you what to look for as you start to analyze usage statistics for your app. One final note: as a developer for the Apple iOS or Android platform, you will typically have access to an administration module that provides the statistics we will be talking through in the next part of the chapter.

One of the first questions you need to answer about your mobile application is, "Are people finding and using my app?" If you were to look at the data collected by an operator portal like Apple's App Store or a web analytics package such as Google Analytics, a number of metrics can help learn how successful people have been in finding your app and how much total traffic your app is getting:

- The "Total Downloads" or "Total Visits" metric shows the number of times your app has been downloaded or visited and measures the total interest in your app.
- The "App Users" or "Unique Visitors" metric is the number of unique users of your app and records how many people are using your app after downloading or sticking around to navigate to interior pages of your mobile website or mobile web app.
- The "New Users" or "New Visitors versus Returning Visitors" metric counts the number of new users or visitors over a period of time. It can be useful in gauging the growth of your site or app.

A second question to consider is, "How loyal or engaged are my users?" With this question, we are looking to find out how engaging or "sticky" your app or site is. This is an essential statistic in understanding how your app is performing in eliciting sustained use. Apps are frequently downloaded or accessed, used initially, and then left alone for long periods of time or removed. What we want to see is people downloading and visiting as well as spending significant time using your app or site. These are useful metrics to watch and monitor for user engagement:

- The "Duration" or "Length of Visit" metric tracks the average amount of time a user spends in your application or site.
- The "Depth of Visit" metric records the number of screens or pages viewed by a user on average compared to the number of visits. It shows the true engagement a user has with your app or site.

A third and final question to ask is, "What are the primary or most popular actions people are performing from within my app?" Common metrics for answering this question include views of screens or pages as well as top visited screens or pages. While analyzing this series of metrics, we are hoping to get a picture of the user's navigation pathways through our app. We are also trying to evaluate what types of actions are worth creating and supporting with this question. Some useful metrics to use here include the following:

- The "Screen Views" or "Pageviews" metric will measure the unique views or actions that a user took when moving through your app.
- The "Top Content" will show the most popular screens or pages, which can help you determine the most useful features to support within your app or site.
- The "Entrance Paths" metric will typically show the path of an average or unique user and the complete recorded actions that user took when moving through your app or site.

REVIEW PROVIDERS' RESOURCES

Having discussed the primary metrics you should consider, a quick survey of some analytics sources and statistics providers for mobile applications will help get you started. First, you have the internal analytics provided by the specific app marketplace or app store. Many of the metrics identified earlier will be recorded by these distribution channels and associated with your developer account. For example, as a registered Apple developer, you would be able to log in to your account and pick through the app store statistics like total downloads to measure the popularity of your app. These internal analytics are useful but can lack the depth of stats you might need to gain larger insights into your app or site development or redesign.

With the cursory nature of the default app marketplace and app store analytics in mind, a number of mobile analytics software companies have emerged to offer more detailed statistics. Bango Mobile Analytics (http://bango.com/mobileanalytics/) is a paid service that enables you to track unique users and their navigation paths with ease. Along the same lines, Admob, a leading mobile advertising service, is an option as they have a mobile analytics service in place (http://www.admob.com/). Flurry (http://www.flurry.com/product/analytics/index.html) is a free mobile analytics service with benchmarks and complete usage data that works across all of the major mobile platforms. And, finally, we have Google Analytics (http://www.google

.com/analytics/), a free statistical service more commonly known for web analytics. Google Analytics also has a mobile application statistics component available at http://code.google.com/mobile/analytics/docs/. (Note: Google Analytics for mobile is a bit more involved to implement and requires you to add tracking code to your mobile app.)

▶9

DEVELOPING TRENDS

- Expect New Apps
- Expect New Resources for Apps

> The only prevailing wisdom in mobile is that there is no prevailing wisdom.
>
> —Daniel Appelquist, one of the W3C Mobile Web chairs

With a nod to Daniel Appelquist's (un)certainty about mobile as a shifting platform, I'm going to put on my futurist goggles for this chapter and work through what some of the developing trends for mobile will be in the coming months. My choice of "months" as the expression for time is deliberate and measured. Mobile and its associated development platforms are changing at an incredible pace. Even with this uncertainty and rapid pace, it is possible to pick out some of the major trends for mobile. Here are some of the possibilities.

EXPECT NEW APPS

HTML5: Native Apps versus Mobile Web Apps

HTML5, a suite of technologies that includes enhanced CSS3 capabilities, new HTML tags, and JavaScript APIs, is making inroads into the spaces only native applications could create just a short year ago. Advanced elements of HTML5—use of geolocation capabilities, access to device file systems, offline storage, drawing of graphics with the canvas element, and so forth—can and will compete with native application functionality. The "write once, run anywhere" dream for mobile applications is becoming closer to reality. As HTML5 matures, expect to see developers picking and choosing when they need to build native applications if an HTML5 mobile web application can perform the same tasks. With budgets shrinking, questions about investing in specific programming languages (e.g., Java, Objective C, C++) for specific mobile platforms will start to be asked. Native apps will still have a place, but mobile web apps built with HTML5 will become a viable alternative.

Rise of the Digital Wallet and Micropayments

Mobile devices will continue to develop as proxies for physical credit cards and paper receipts. Devices with airline itinerary bar codes or lines of credit within Apple's iTunes service are just a few current examples. Square (http://squareup.com/), a combination of hardware and software that allows vendors to accept credit card payments using mobile devices, is quickly moving mobile in this direction. Along similar lines, Google Wallet (http://www.google.com/wallet/) is an app that actually makes your phone your wallet by storing digital versions of your existing, physical credit cards on your phone. More companies will realize the opportunity to apply mobile devices to e-commerce settings and introduce micropayments, mobile transactions that involve a small sum of money. Want to upgrade that app to the premium version? A micropayment transaction will be in place to allow this.

Further Resources

To learn how you might create mobile payment services, see the working project titled "Implement a Mobile Payment Service with Google Wallet and Near Field Communication at Your Library" in Joe Murphy's *Location-Aware Services and QR Codes for Libraries* (THE TECH SET #13).

Location-Based Apps

In mobile settings, context is king. One of the benefits of mobile devices is that they are on hand, connected, and able to be positioned in time and space using GPS. More and more apps will become location aware. Facebook Places, Foursquare, and even Twitter are bringing location-based apps into the mainstream. Many phones and digital cameras now automatically "geotag" pictures. As the interest in place grows, devices and services that use location as a metric for interest will grow as well. Contextual, historical walking tours of the National Mall in Washington, DC, with the help of an app on your smartphone are being developed. What could be next?

▶ EXPECT NEW RESOURCES FOR APPS

Mobile SEO and Optimizing Apps for "Findability"

With the increasingly divided distribution channels and the sheer volume of apps produced, the need to make sure your project will be found and used will be imperative. New methods of coding and "marking up" apps for retrieval by search engines and app store searches will be introduced and tested. Start-ups and enterprising individuals will offer solutions for making your app findable as a "signal in the noise."

New Marketing and Analytics for Mobile Applications and Sites

Along similar lines, the need for marketing apps and analyzing their use will continue to grow. Learning to distinguish your app in a crowded field will be essential. Emerging methods of promotion like app store product reviews and user ratings will be pushed (and even gamed). Mobile analytics as an industry will mature quickly in response to the limited data provided by distribution channels and app marketplaces.

Rise of Multiple App Stores and Marketplaces

The fragmentations of devices, browsers, and computer platforms will lead to multiple, disparate distribution channels. We are already seeing this happen. Google has a Chrome Web Store (http://chrome.google.com/webstore/) for specific browser apps. Apple has a "Web App" app store (http://www.apple.com/webapps/) that competes with its original native apps "App Store." As more devices are released, each product has its own catalog of apps in its own proprietary silo. Searching and browsing across these silos will be a challenge. Somebody will have to do what Google did for the web in the mobile setting.

RECOMMENDED READING

This collection of websites, blogs, feeds, and other resources will help you stay on top of the mobile ecosystem and will point you to useful development resources as you start to put the ideas from this book into practice.

CURRENT AWARENESS SOURCES

Blogs

In terms of current awareness and breaking news in the mobile field, a good place to start is with the major technology news sources.

Mashable (blog). 2011. Accessed November 7. http://mashable.com/.

Mashable is a news aggregator centered around social media happenings, digital culture, and Internet history. Sections on mobile and app development are particularly useful for breaking news about the mobile platform.

ReadWriteWeb (blog). 2011. Accessed November 7. http://www.readwriteweb.com/.

This is another top technology blog that provides essential reading related to technology developments and Internet culture. *ReadWriteWeb* has a dedicated mobile channel available at http://www.readwriteweb.com/mobile/.

TechCrunch (blog). 2011. Accessed November 7. http://techcrunch.com.

This web publication focuses on start-up and web development products. Mobile topics often break here, and monitoring new services announced here provide a forecast of things to come for mobile development. *TechCrunch* has created a mobile section at http://techcrunch.com/mobile/.

Tagwatching

Tagwatching is another useful means of gathering info related to your topic of interest. Tagwatching involves choosing a service or a website, finding a relevant tag, and then watching or monitoring the tag in a feed reader or your web browser. You can combine and specify multiple tags in many of these services to refine the feeds to your particular interest. Here are a few examples to get you started:

http://www.readwriteweb.com/tag/mobile
http://pinboard.in/t:mobile/
http://pinboard.in/t:mobile/t:app/
http://www.delicious.com/tag/mobile+android

► MOBILE DEVELOPMENT RESOURCES AND TOOLS

Mobile development resources are plentiful, but here are a few selections that will be important as you continue learning.

Boopsie. 2012. Accessed January 11. http://www.boopsie2.com/.

If you are interested in outsourcing app developments, check into Boopsie. Boopsie specializes in mobile apps for universities and libraries.

Library Anywhere. 2012. LibraryThing. Accessed January 11. http://www.librarything.com/forlibraries.

Library Thing, like Boopsie, is an excellent company that has experience making mobile work for libraries.

MobileOK Checker. 2012. Accessed January 11. http://validator.w3.org.

Validating and testing your app is essential. The W3C's MobileOK Checker can help you debug your mobile web app and teach you in the process. Best practices related to mobile web apps have been collected by the W3C at http://www.w3.org/TR/mobile-bp/.

MobileTuts+. 2012. Accessed January 11. http://mobile.tutsplus.com/.

If you are looking for tutorials for building native apps and mobile sites, MobileTuts+ is an excellent resource.

Mobjectify. 2012. Accessed January 11. http://www.mobjectify.com/.

If you are looking for a simple tool for prototyping your mobile design, check out Mobjectify.

► BOOKS ABOUT MOBILE DESIGN AND DEVELOPMENT

Fling, Brian. 2009. *Mobile Design and Development*, 1st ed. Sebastopol, CA: O'Reilly Media. Also available online at http://mobiledesign.org/toc.

There are many books available on this topic, but a great resource during my writing was Brian Fling's O'Reilly manual. Not so much focused on the practical, but it is readable, heady, and worth the purchase. Also see Brian Fling's blog at http://pinchzoom.com/blog/ for Brian's latest thoughts on app development and design.

Griffey, Jason. 2011. *Mobile Technology and Libraries*, 1st ed. New York: Neal-Schuman.

Jason Griffey's study of mobile technology in libraries provides some important context for our institutions.

Stark, Jonathan. 2010. *Building Android Apps with HTML, CSS, and JavaScript*, 1st ed. Sebastopol, CA: O'Reilly Media; *Building iPhone Apps with HTML, CSS, and JavaScript: Making App Store Apps without Objective-C or Cocoa*, 1st ed. Sebastopol, CA: O'Reilly Media. Also available online at http://ofps.oreilly.com/titles/9780596805784/.

These step-by-step manuals for Android and iPhone development focus more on the practical code of building mobile web applications than on native applications.

▶ "LIBRARY-CENTRIC" PRESENTATIONS, PAPERS, CONFERENCE PROCEEDINGS, WIKIS

Finally, a number of library resources (presentations, papers, conference proceedings, wikis, etc.) related to mobile development and design are available.

Bridges, L., Hannah Gascho Rempel, and K. Griggs. 2010. "Making the Case for a Fully Mobile Library Web Site: From Floor Maps to the Catalog." *Reference Services Review* 38, no. 2. http://ir.library.oregonstate.edu/xmlui/handle/1957/16437.

This solid article talks through the conceptualization of mobile services and how to build a business case for mobile at your library.

California Digital Library, Mobile Device User Research. 2010. *Resources*. California Digital Library. Last modified September 9. https://confluence.ucop.edu/display/CMDUR/Resources.

This annotated bibliography contains a series of mobile statistical studies conducted by a leading digital technology group. It is useful in distilling the current statistics related to device adoption and usage analytics of the mobile platform.

Farkas, Meredith. 2011. "M-Libraries." In *Library Success: A Best Practices Wiki*. Last modified November 30. http://libsuccess.org/index.php?title=M-Libraries.

"M-Libraries" is compiled by library editors and lists the "best of breed" mobile solutions for all kinds of library organizations and mobile projects.

Handheld Librarian (conference). 2011. http://www.handheldlibrarian.org/archives/.

Handheld Librarian is an annual online conference for librarians to discuss the emergence of the mobile platform and how libraries are building new services to address mobile learning. Past conferences are archived here.

Kroski, Ellyssa. 2008. "Libraries to Go: Mobile Tech in Libraries." SlideShare presentation. November 3. http://www.slideshare.net/ellyssa/libraries-to-go-mobile-tech-in-libraries-presentation.

Kroski provides a good overview of getting started with mobile development in the library setting.

Sierra, T. 2010. "Opportunities for Mobile Enhanced Library Services and Collections." Presented at the Johns Hopkins University Libraries Assembly, May 21, 2010, Baltimore, MD. http://www.lib.ncsu.edu/staff/tsierra/presentations/jhu-assembly-2010.ppt.

This talk was given by a library developer with good insight into how the mobile platform can be applied to library settings.

Woodbury, David, and Jason Casden. 2010. "Library in Your Pocket: Strategies & Techniques for Developing Successful Mobile Services." EDUCAUSE Seminar. January 20. http://www.educause.edu/Resources/LibraryinYourPocketStrategiesa/195003.

This web resource from EDUCAUSE looks at mobile possibilities for higher education and learning organizations like libraries.

INDEX

Page numbers followed by the letter "f" indicate figures.

ABOUT THE AUTHOR

Jason A. Clark is Head of Digital Access and Web Services at Montana State University Libraries where he builds digital library applications and sets digital content strategies. He writes and presents on a broad range of topics, including mobile design and development, web services and mashups, metadata and digitization, JavaScript, interface design, and application development. Before coming to MSU in 2005, Jason became interested in the intersection between libraries and technology while working as a web developer for the Division of Information Technology at the University of Wisconsin. After two years, he moved on to the web services department at Williams College Libraries. When he doesn't have metadata on the brain, Jason likes to hike the mountains of Montana with his wife, Jennifer, their daughter, Piper, and their dog, Oakley. You can find Jason online by following him on Twitter at twitter.com/jaclark, catching up with him on Facebook at facebook.com/jasclark, or checking out his occasional thoughts and code samples on his site at http://www.jasonclark.info.